MAN MEDICINE and SPACE

A Manifesto for the Millennium

Dr.Michael Martin-Smith

Writers Club Press
San Jose New York Lincoln Shanghai

MAN MEDICINE and SPACE
Michael H Martin-Smith

All Rights Reserved © 2000 by Dr Michael Martin-Smith

No part of this book may be reproduced or transmitted in any form or by any means, graphic, electronic, or mechanical, including photocopying, recording, taping, or by any information storage retrieval system, without the permission in writing from the publisher.

Writers Club Press
an imprint of iUniverse.com, Inc.

For information address:
iUniverse.com, Inc.
5220 S 16th, Ste. 200
Lincoln, NE 68512
www.iuniverse.com

ISBN: 0-595-14808-5

Printed in the United States of America

Dedication

This book is dedicated to my dear wife, Irene, whose help and comments have been invaluable during the past few years, and to our children Leo and Alexander, for whose generation this book, with its central proposition, is intended.

I would like to take this opportunity to express my appreciation of the late Dr. Hermann Oberth, whose work has helped to make the development of space an engineering reality, and whose dream of a popular movement of enthusiasm for the realization of the benefits of space for all humankind came so near to reality in his own lifetime. My thanks also go to Arthur C.Clarke for planting the dream in the minds of so many of my generation, and to the late Professor Gerard K. O'Neill for giving the dream flesh and bones.

Dr.Michael Martin-Smith
September, 2000

Introduction

This book is the result of a lifelong interest in space and the possibilities it offers Humanity.

My interest in the subject of space began when as a small boy I became curious about such age-old questions as "Are there other worlds like ours?" and "What are their inhabitants like?" Curiosity about undiscovered lands and people has led to great voyages of exploration. Indeed, curiosity has been regarded as a defining characteristic of the human race.

As I learned gradually through childhood, monsters, demons, fairies and the like have been banished from the Earth, by advances in knowledge. Fantasies of this sort were therefore firmly banished to the Sky, where they would remain safe from our prying eyes. The planets, were remote, mysterious, and suitable homes for fabulous beasts; Venus was tropical jungle, and Mars the land of reptilian or insect-like men. I remember receiving from my father a small East European publication entitled 'Travel to Distant Worlds' in which the possibility of rocket-powered flight to other planets was canvassed. I read with excitement, little dreaming how soon these worlds would be visited.

1957 will be remembered by almost everyone-in this year the first artificial Earth satellite burst upon the consciousness of millions. For me, as a young schoolboy , this was a sign that the lurid jet-blazing rockets of science fiction could be the pathways to tomorrow. To myself, and, many others, the future role of Humanity would never look quite the same again.

My interest broadened into the realization that there was a growing economic and cultural effect on human society around the world from the activities of Humankind in space.

I discovered that many writers and pioneers had considered, as I had, the age-old questions, and that they have felt that the answers, would have a great bearing on the role of Humanity in the Universe.

Since my childhood , our operations in space have broadened out and come to involve financial and economic developments, which are beginning to affect the nations. In the future, the effects of space development on human societies will become more marked, and, apart from providing us with a much needed frontier, will contribute to the solution of many serious problems on Earth.

In writing this book, I am hoping to provide a brief, and easily accessible description of where space exploration and development fit into human evolution. I believe that I shall be offering to the reader a much more positive view of the future than he or she may be used to. I make no apologies for the partisan nature of this work. I have not set out to write a treatise on astronautical engineering, bio-medicine, or space history. The work contains elements of all these , and more-but, most importantly, it is a call for action, and sets out a new ideology for the next millennium.

In order to achieve this, the first chapters will outline current ideas of the origins of the Universe, from an initial "Big Bang", and show how from an early stage, there seems to be a cosmic order leading to higher states of order and consciousness. Here I describe how there appears to be a purpose driving Life to greater consciousness and adaptability, and that migration into space is an integral part of this process.

Next, I describe how human mythology and psychology is concurrent with this idea; in describing ancient mythologies of flight, and the search for "Perfection" among gods in the skies, I believe that this demonstrates a basic, deep-seated drive of the human psyche towards space. It follows, that the Space Age is not an unnatural growth of political rivalry, but is the realization of the evolutionary drive itself. In so far as evolution is creative, and the manifestation of a higher purpose, our activities in space are an integral part of that process.

Later chapters describe the great explorations of earlier generations, as these provide parallels for the settlement of the Solar System. I go on to describe the origins of spaceflight, some of its history, and some of the discoveries made in this first generation of the Space Age. As a doctor, I have concentrated on the adaptation of Man in space, and how the problems encountered will be overcome, as the revolution in space proceeds. Nothing has been discovered as yet which would discriminate against women in space-indeed the evidence suggests that all the divisions of Humanity are likely to pale before the magnitude of the tasks ahead.

Later chapters describe, firstly, the coming revolution in space, and the new processes which will be opened up. As medical pharmaceuticals form an important part of this stage, I have described them at some length. Many people are genuinely excited by this aspect, as it will bring real and indisputable benefits to millions of tax-payers.

The role of space in solving the world's energy, and resources crises is described at length, since many people have criticised the expenditure of resources on "space junkets" when so many serious problems remain unresolved.

It is my contention that solutions to these problems will become uniquely available from the results of space activities, and that they provide for a humane future. In view of the widespread feelings of despair and futility I have encountered, I have met this argument head on. It is worth pointing out that the *annual* global cost of the illegal trade in abused "substances" is $300-400,000,000,000,000-which dwarfs anything I or anyone else has suggested doing in space! The cost of puerile ethnic warfare, and its sustaining armaments trade, is something else again.

Assuming these developments, and that we all find our way safely through the very turbulent half century or so ahead of us, my last chapters attempt to show how a space based civilization could take shape, and where it could lead. For it is my belief that our destiny lies in space, and that if we can use space to help solve our ecological problems over the next 50-100 years, we have the chance of a diverse and prosperous future for our species. The price of failure would not only be dire-it has even been graphically demonstrated on the windswept desolation of Easter Island, and, more recently, in the bombardment of Jupiter by Comet Shoemaker-Levy 9.

Unrestricted by limits of room and energy, future generations will be able to explore and diversify, and write new pages in the Book of History.

Before we can consider our destination as a species, we must turn to our origins; where better to start than at the Beginning?

Contents

Dedication iii
Introduction v

Chapter 1 1
Chapter 2 25
Chapter 3 44
Chapter 4 67
Chapter 5 90
Chapter 6 108
Chapter 7 135
Chapter 8 153
Chapter 9 194
Chapter 10 216
Chapter 11 229

Epilogue 237
Bibliograph and Sources 239

Chapter 1

Evolution and Origins-"From dust are ye come......"

As the Millennium approaches, Humanity is arriving at a significant stage in its long attempt to understand the Universe. For many millennia, such tasks have fallen to religion, mysticism, and philosophical speculation. In past centuries, and especially in this one, the process of observation of Nature, with derivation of her underlying principles-the process known as natural philosophy, but now called science-has extended our study into distances and timescales scarcely conceivable to our forbears. In so doing, we have cast new, sometimes cold, light on to our traditional beliefs and values. Some indeed have held science responsible for a loss of meaning in human culture, and ultimately destructive. One of my objectives is to show how science and technology, so far from being destructive of humanity and its role in the universe, shows the way to a more firmly grounded view than any that have so far inspired human minds. In this new view of the Universe, we are speaking not so much of a revealed blueprint, but rather the opportunity to build our own purpose, based on a clearer view of our origins, development, and of our universal setting. It is not yet possible to give a complete account of the origins of the Universe, Solar System, or life on Earth. However, sufficient advances have been made to give us a general outline.

The most widely accepted theory of the origin of the observable Universe is known as the "Big Bang" theory; this theory states that, at a definite time in the past, reckoned as about 12-15 billion years ago, the Universe was exceedingly small, dense, energetic and hot. For reasons not fully understood, this primaeval ball of fire, or Cosmic Egg, with its small contaminant of matter, inflated out of a virtual vacuum of quantum energy, thereby undergoing a titanic explosion. Even now, the galaxies of this Universe, embedded in the very matrix of space-time itself, are being blown apart from each other, and the Universe is still cooling down from the unimaginable heat of the explosion. The theory is not proven, but, over the last 60-70 years, astronomers have gathered an impressive body of observations to support it.

Firstly, in the 1920s the American astronomer Edwin P. Hubble discovered that the light emitted by some galaxies was altered; its wavelength was shifted towards the red end of the spectrum. You will have noticed that as a train approaches a level crossing at high speed, the whistle sounds a higher pitched note as it approaches you, while, as it moves away, the pitch drops. This is because the sound waves, travelling at the same speed, are compressed as the train approaches, and stretched as it retreats. The same process occurs with light and other electromagnetic waves (e.g.radio, ultra-violet, or Xrays). For a higher pitch the light waves become shorter-bluer, while a reduction in pitch corresponds to a reddening. Hubble noticed that many galaxies emitted light which was definitely redder than it should be, and that this could be seen no matter in which direction the Universe is observed. This alteration in light wavelength became known as the Cosmological Red shift. The next step was the discovery by Hubble and others that the degree of red shift depended directly on the distance of the galaxies. This law, Hubble's Law, is one of the most basic of modern astronomy, and has stood the test of over 50 years. Extremely faint galaxies and quasars have been detected from the ground, and more recently from the Hubble Space Telescope, at distances of 12 billion light years ; from their red shifts, it has been calculated that they are receding from us at over 90% of the speed of light. It follows that since they are over 12 billion light years away, the light left them 12 billion years ago. As Einstein showed that no material can travel faster than the speed of light, a red shift equivalent to 100% of light speed would be unobservable, and would be a contemporary of the Big Bang. Thus, our observations of the quasars and galaxies reach back over 90% of the time (distance) to the Big Bang itself.

This expansion has been observed in all directions and is believed to be similar no matter from which viewpoint observations are made; any observer would observe an expanding Universe with him/her/it self at the apparent centre. For example, imagine a childs balloon, with many dots drawn on the surface. These dots represent galaxies and quasars; as the balloon is inflated, a miniature observer on any of the dots sees the others rushing away , with the farthest ones seeming to move the fastest. If the observer is clever, he might be able to deduce the size of the balloon, and when the inflation began.

This, then, is the big bang theory; a Universe which began its expansion some 10-20 billion years ago from an initial fireball. It is expected that, by the new millennium, the Hubble Space Telescope will have refined the timing to an accuracy of ± 10%. The red shift strongly implies the expansion, but the evidence for the fireball is also strong. In 1965 Penzias and Wilson discovered a universal background radio noise when experimenting with new communications equipment. In whatever direction they pointed their antenna, they picked up a low-level background micro-wave hiss, such as might be emitted from a body cooled to only 2.7K. The average living room is heated to about 293K, or 20C. This radiation is now recognized, from rocket-borne and satellite measurements, to be "black body" radiation, and is proportional to the energy content of the emitting body. A good parallel is a red hot lump of metal- because it is hot, it emits red light; if heated further, it becomes white hot. The wavelength becomes shorter as the energy radiated increases. As the metal cools, it loses its glow, and radiates only heat which, having lower energy than light, is called *Infra*red, and is, invisible.

Just as gas cools down when it expands, so did the hot dense Universe during its long expansion-indeed, it is still doing so, billions of years later: it is now so cool and diffuse that the background radiation is only 2.7K.

More recently, in 1992, the Cosmic Background Explorer satellite (C.O.B.E), refined the "fit" of the curve of the background radiation to a typical black-body emission of 2.735.K, from 2 billion measurements over 2 years to an accuracy of 1 part in 30 million and, indeed did something even more. One remaining problem with the scenario was that, out of such an apparently "smooth" fireball was born a very "lumpy", uneven Universe characterized by dense clusters of Galaxies embedded in oceans of pure vacuum. Surely, at about the time of the separation of matter from background radiation, after the Big Bang, there should be fluctuations in the distribution of energy as a fossil of irregularities in the original fireball, from which the galaxies were to condense. After 2 exhaustive surveys of the whole sky, and over 2 billion observations, meticulously scanned by image processing computers, C.O.B.E produced the famous ripples of energy in the distribution of the background radiation, accurate to 1 part in 30 millions. Not quite the "glimpse of the Mind of God" , but surely as near to verification of the Big Bang theory as we can reasonably expect. However, particle physicists, with their cyclotrons, are now achieving, conditions similar to those within one billionth of a second after the Big Bang, mirroring the calculations of the cosmologists.

From these 3 discoveries-red shift, micro-wave background radiation and the C.O.B.E ripples-it has proved possible to say a lot about the early history of our Universe. Up to the first 700,000 years after the beginning, radiation and matter were in flux, with only primitive short-lived sub-atomic particles having an independent existence initially, with atomic nuclei of hydrogen and deuterium towards the end of this period. Indeed, the ratio of hydrogen to deuterium, and of helium to hydrogen in the Universe is dependent on the rate of cooling and expansion of the early Universe, and is another verification of the Big Bang calculations. Only after 700,000 years or so could energy and matter "decouple"; matter condensed out of a homogeneous flux of matter and radiation in which packets (quanta) of radiation outnumbered particles of matter by 1,000 millions to one. In the minutes after the beginning, the Universe was energy, with a trace contamination of matter to the extent of 1 part in a billion, whereas today the Universe is dominated by matter, with energy in the cosmic background. Recently the Hubble Space Telescope has seen an absence of galaxies in the first 1-2 billion years between the COBE era, and the appearance of the first small galaxies, confirming that there has been change in the history of the universe, and that there was a pre-galactic era. In a few years, it is possible that Hubble will actually see a fireball of radiation 10-20 billion light years away, only 700,000 years after the Big Bang itself-indeed, C.O.B.E, has already done so; further than that, it is doubtful if the eye can see. Even as early as this, it is possible to say that the odds against a universe suitable for civilized life were remarkably high; if the radiation to matter ratio were greater, and the radiation pressure driving the matter outward, were greater than the actual value by more than 1 part in a billion, then there would be no condensation of matter into stars and galaxies. Instead there would have been a dispersed cloud of tenuous gas. If, matter had been one part in a billion more prevalent, gravitational attraction would have ensured collapse within a few hundred thousand years, so that the possibilities for life would have been extinguished at birth. There are many such crucial situations in the history of the Universe, as a result of which the emergence of conscious life appears less like an accident, and more like a conspiracy!

Birth of Stars and Planets

However, we know that stars' and galaxies have risen and fallen, and have a good idea of stars life-cycles. Stars are of many different sizes, colours, and ages; it is accepted that the mass (amount of matter) in a star determines the size, colour, energy radiation, life-span and fate according to predictable laws. Although stars are much longer lived than humans, there is a huge population to sample, so that one can deduce the lifecycle of stars by seeing many individuals, at different stages, and apply our knowledge of nuclear and plasma physics and the laws of thermodynamics. In regions like the Great Nebula of Orion, and the Pleiades Cluster, it is possible to observe vast clouds of interstellar dust grains and gases, and using radio and infra red astronomy, detect chemical elements and compounds within them. These turn out to be substances easily matched by spectroscopy with chemicals found on Earth, and the results are astonishing. Water, hydrogen, hydroxyl, ammonia, methane, formaldehyde, cyanogen, ethyl alcohol, cellulose, urea, amino acids and chlorophyll-like organic rings have been detected by examining the radiation absorbed and emitted from these clouds, which are well known to be the birthplaces of stars and planetary systems. Indeed, since the 1960s, several hundred molecular species have been found. More recently, the instruments of Europe's space probe Giotto in 1986 have found that the nucleus of Halley's Comet is composed of 10% organic chemicals. Stanley Miller, in 1952, exposed mixtures of ammonia, methane and water to UV light and electrical discharges and succeeded in generating 14 of the 20 amino acids essential to life on Earth, as well as carbohydrates, within a week. Further experiments went on to produce purines and pyrimidines, the base blocks of nucleic acids. Thus the seeds of life are to be found in stardust. Professors Sir Fred Hoyle and Chandra Wickramasinghe have suggested, amid great controversy, that some interstellar dust grains might even be bacteria. Certainly, the size and absorption characteristics of some of them can be matched with cellulose, a key component of cell wall membranes. Sample return missions to comets and carbonaceous asteroids are likely to settle this question.

After the Big Bang, the Universe consisted of hydrogen, heavy hydrogen, and helium; the theory of star formation tells us that, there were places where concentrations of hydrogen reached a high mass and density. These gas clouds would condense into denser clouds , in which the concentrations of hydrogen and deuterium would rise to the point that gravitational forces would act on the individual atoms, causing them to fall towards the centre with high energy, and the cloud would heat up. In time great temperatures and pressures would build up, and infra-red radiation would be emitted. Such a hot gas cloud, emitting infra-red radiation through a mantle of dust, is called a protostar, and several have now been detected in the dust clouds of Orion, 1500 light years away. The speed of gravitational collapse is determined principally by the mass of the gas cloud, or the number of atoms, and their density in the original cloud. If the gas cloud heats up too quickly, the incipient star can be blown apart by the rising tide of radiation, so that the moderating presence of dust particles allows more hydrogen to accumulate before the temperature rises to the critical point. At this temperature, about 20 million degrees C, the hydrogen and deuterium atoms collide, and in some cases stick together. This process is called nuclear fusion and is very difficult. On Earth the only way is to use an atomic bomb as a trigger; the heat provided by the Uranium bomb trigger is a "thermonuclear" reaction and the fusion of hydrogen atoms into helium leaves some surplus mass, which by Einsteins $E=mc^2$ equation, yields a formidable amount of "free energy"-the hydrogen bomb. Over the last half century , scientists have tried producing fusion energy by heating hydrogen/deuterium mixtures to 20 million degrees without the uranium trigger. The two main approaches have involved either heating the hydrogen with a high electromagnetic field and containing the resulting plasma in a doughnut shaped magnetic "bottle" (e.g. Tokamak, J.E.T etc.,) or impacting a pellet of frozen hydrogen/deuterium with high energy lasers (Lawrence Livermore Laboratory, California). The principal difficulties have centred around the instability of the plasma, which make it difficult to achieve the required 20 million C x 1 second x $1*10^{14}$ particles/cubic millimetre. The best results have produced demonstrable nuclear fusion, but have yet to yield more energy from fusion than has been used to heat the plasma. The stars employ the massed gravitational energy of $1*10^{57}$ particles to create the necessary heat and pressure. In fusion, two atomic nuclei of the simplest element-hydrogen, and its heavier isotope, deuterium, stick together to form helium. There are 92 natural elements in chemistry; hydrogen is element 1, with one proton in the nucleus, balanced by one orbiting electron of opposite electrical charge, while uranium is element 92-with 92 positively charged protons, 146 electrically neutral neutrons, and 92 negatively charged electrons. The neutrons provide relative stability by providing a strong nuclear force without a disrupting electric charge. Without the neutrons mass and strong nuclear attractive force, the positively charged protons would repel each other; whilst in the larger nuclei the strong nuclear force, which operates only over very short distances, begins to break down, with the result that the heavier elements fall apart in nuclear fission. Deuterium is hydrogen with a neutron as well as a proton in its nucleus. The exact ratios of the forces to which I have referred-strong nuclear, electromagnetic and gravitational-are all exquisitely balanced to allow chemistry which is stable, but reactive enough to allow evolution and Life. The energy produced by the fusion of hydrogen into helium ($E=mc^2$) is due to the fact that a helium atom weighs only 99.5% of the sum of its constituent atoms, and is seen as starlight, or sun light and warmth in the case of our Sun. As we have seen, controlled nuclear fusion is a Holy Grail among scientists all over the world, both for its clean, inexhaustible energy, and its possibilities in advanced propulsion systems-but, as we shall see , Natures own fusion reactors-the stars-have done the work for us, and with space based industry, we have only to collect it. The Sun has provided fusion derived energy for the Earth for 4.6 billion years , and is expected to continue for another 6 billion years. But what happens when all the hydrogen has fused into helium, and there is no more "fuel" left?

This depends upon the initial mass of the star as it forms from the condensing gas and dust. The more massive the star, the more gravity predominates over the pressure of radiation, which results in a more rapid consumption of "fuel", since gravity, being stronger heats up the core to a higher temperature. Because it is hotter, its radiation predominates in the shorter, energetic wavelengths. Shorter wavelengths are more energetic because there are more waves to the metre. Thus very massive stars are blue-white to the eye, and flourish for only tens or hundreds of millions of years-Vega, Rasalhague, and Rigel are typical Blue Giants; whereas small cooler stars are red and long-lived, with life-spans running into the trillions of years-Red Dwarfs Barnards Star and Wolf 756. The fact that Barnards Star-is only 6 light years away and is never seen the public, while Rigel at 900 light years is known to everyone who has glanced at Orion on a clear winter's night illustrates my point. As the stars reach the end of their hydrogen reserves, gravity asserts itself over the waning radiation pressure, and helium is pulled in towards the core, where it becomes hotter. If sufficiently hot, helium fuses into carbon, with a release of energy, and expansion of the star. The star becomes unstable, with alternating cycles of expansion and contraction as larger atomic nuclei are created. Sometimes, rings of the star's outer layer are blown onto space-these are seen as nebulae from Earth. These "smoke rings" of outer gas and dust are recycled into the interstellar medium, to become part of future generations of stars. Perhaps the best known and loveliest example is the "Ring Nebula" in the summer constellation of Lyra. These unstable stars are "long period variables" because they are of regularly variable brightness, with cycles ranging from 150-700 days, and are often followed by amateur astronomers. During this stellar dilatation, the star is called a Red Giant. In more massive stars, this expansion is much more rapid, so that a point is reached where the heaviest element formed is iron-at which time the star faces a crisis. Up to iron the fusion of heavier atomic nuclei has resulted in an output of energy, albeit for decreasing lengths of time; however, once iron is reached, the next fusion requires an input of energy. Gravity, now gives a whoop of triumph and, in seconds, the outer layers of the star implode towards the core. This results in a titanic burst of energy, in the form of neutrinos and a massive shockwave. The shock wave rips through the outer layers, causing a rapid succession of nuclear fusions right up the table of elements up to and including uranium, like a musical crescendo! The outer layers are ripped out into space with a cataclysmic force which is detectable as a supernova. This outer layer, with its heavy elements, is scattered over interstellar space, and is recycled into the formation of stars and solar systems. All the elements on Earth-for example nickel, gold, platinum, and iron-are thus recycled from the death-throes of some long dead star: indeed, the trace radioactivity in some meteorites has shown clearly that some of these predate our solar system, and that our system was called into being as a result of a supernova shockwave compressing a gas cloud , thus triggering star formation. Furthermore, since our blood contains haemoglobin as the vital oxygen bearing pigment, and since its molecular structure contains iron atoms, we can claim to be children of the stars, formed from their ashes. It is even possible that the iron in our veins once nourished the denizens of some long dead alien race, whose achievements are lost forever.

We have considered the outer layers of various stars, but what of the remaining core. This turns out to depend on the mass of the stars as they approach their Gotterdammerüng. In stars from 0.7 to 1.4 solar masses, the core condenses to a hot core of superheavy matter the size of a medium planet. Gravity squeezes the atoms of the core so closely that the electrons are squeezed right down to the core, so that the atoms lose their electron shells. This degenerate matter is very dense, so that a teaspoonful will weigh several tons. The white-hot stellar remnant shines on as a "white dwarf" and over aeons burns itself out into a cinder, known as a black dwarf-these have not been seen! White dwarfs, however, are well known-indeed, the first known example orbits Sirius, 8 light years away, and was detected in 1862.

The fate of more massive stars, after the outer layers have been blasted off, depends on the amount of mass left behind; between 1.4 and 3 solar masses results in a more extreme triumph of gravity-the orbiting atomic electrons are squeezed right out of existence into the core, overcoming their mutual electrostatic repulsion, and the atomic nuclei are converted into a sea of undifferentiated neutrons-a neutron star. A teaspoon of matter now weighs hundreds of millions of tons, the "star" being only a dozen miles across. Usually, these neutron stars rotate at a fantastic rate-all stars rotate, and, as they shrink, the speed of rotation increases, in the same way that a ballet dancer can spin faster by lowering her arms to her sides. Over the course of history several of these brilliant supernovae have flared up, outshone all other stars and then faded without trace. Or have they....?

The celebrated Supernova of 1054 outshone Venus in the day sky for weeks, and was recorded in China, and Constantinople, and then faded out of sight-until the late 1960s, when Jocelyn Bell-Burnell and Anthony Hewish at Cambridge discovered a new class of rapidly rotating object, which appeared to pulsate with all the rapidity and regularity of a quartz clock. It was realized that this object was one of the long-theorized neutron stars, and was clearly embedded with a faint, irregular patch of light in the constellation of Taurus, known as the Crab Nebula. Calculations of the rate of expansion soon revealed that the "Crab" originated 900-1000 years ago, while a gradual slowing down of the object's fantastic spin rate of 30 times per second led back to a similar date. It was evident that the mystery "pulsating" star-now dubbed a pulsar-is the remnant of the famous historical supernova. Several hundred of these remarkable objects are now known, and some have now been imaged by the Hubble Space Telescope.

Beyond 3 solar masses, the collapse of the dying star is even more complete-it becomes so dense that gravity will not even allow light to escape its surface. With 3 or more solar masses concentrated into a region scarcely bigger than my city of Hull this is the final annihilation which we call a "black hole". It is now believed by some that the undetectable matter is squirted out of this Universe into another space-time continuum-and that our disappearing black hole is some-one else's Big Bang. Black holes cannot be directly seen, but can be detected by their effects on nearby companion stars, or, if large enough, by their surrounding galaxies. There are several likely candidates for black hole status, of which the nearest and best claimant is an Xray emitting object known as Cygnus X1. Until 1987, this account of supernovae was largely theoretical since no nearby supernova has erupted since the invention of the telescope. However, in 1987 a new, bright, star was seen in one of our neighbouring galaxies, the Large Magellanic Cloud-a next door neighbour 169,000 light years away. Its evolution has been followed by observers on Earth, the Mir space station, the Space Shuttle, and with the Hubble Space Telescope with great excitement-the observations confirm, in all essentials, the outline described above. We can thus see, in the account of the life, death, and recycling of stars and interstellar dust, a cosmic ecology, the next phase of which is the formation of solar systems.

The formation of planetary systems has traditionally been accounted for by two main theories. Firstly, Kant and Laplace in the 18th century supposed that the planets accreted out of swirling clouds as they circled the newborn Sun, and that successively larger particles, pebbles and eventually mountain sized boulders and snowballs clumped together to build the planets. If this were true, planetary systems ought to be widespread.

The second theory, that of Jeans and Eddington in the 1930's, was that another star, randomly wandering into the region of the Sun, drew off by gravitational pull an elongated cloud of matter which broke into fragments which we now recognize as the planets-this would be an event of great improbability.

The evidence from meteorites, and cratering on planetary surfaces, as well as from lunar, Martian and Venusian drill samples now confirms that the entire solar system formed at the same time, about 4,550 billion years ago, and that, in the earliest stages, the planets all formed together by the accretion of "planetesimals"-as Kant and Laplace supposed. The asteroids and comets are now understood to be primordial matter that missed being "accreted" into any embryonic planet. Indeed some meteoritic material provides evidence that the birth of the solar system from a protostar embedded in an interstellar cloud was triggered by a nearby supernova explosion. In this connection, the discovery of large amounts of organic chemicals in the comets, and asteroid belt, is of great interest. Meteorites indeed fall into 3 classes-nickel/iron, stony, and carbonaceous chondrites-the rarest of all. Examples from Antarctica and the extreme tip of South America-the Murchison and Allende meteorites-have been found to contain traces of organic substances in their depths, including porphyrin rings-a building block of haemoglobin and chlorophyll.

The dust clouds in the Orion Nebula have revealed many infra-red sources, and some early stars of the protostar type. Indeed, some have been reported as less than 1,000 years old, with the clear inference that the dust clouds would condense into planetary systems. These findings, from the I.R.A.S (Infrared Astronomy) satellite in 1983, have since been extended by the repaired Hubble Space Telescope in 1994. Dozens of infant stars complete with surrounding accretion shells have been documented. Thus there is little doubt that new planetary systems are actually being born in the nurseries of Orion. The I.R.A.S satellite also detected, in May 1983, a ring of small dust particles in orbit around the brilliant blue-white star Vega, at 28 light years from the Earth: for summer campers, Vega is the brilliant apex of the "Summer Triangle" seen almost overhead in the skies of August and September. This almost certainly represents the birth of a Vegan planetary system: sadly, with a life expectancy of 300 million years, it is unlikely that Vega will nourish any native civilizations. Similar particulate rings have been found orbiting Fomalhaut at 22 Light years, and Beta Pictoris at 50 Light years. In late 1997, UK astronomers, using newly developed adaptive optical telescopes, confirmed that these dust clouds together with one orbiting the star Epsilon Eridani, had gaps, swept out by early planetary masses. IRAS examined 8000 other sources in its 18 months of work, and more discoveries can be expected. This is already very strong evidence for the Kant-Laplace theory, but the final step, actual detection of an extrasolar planet, is becoming feasible in the next few years.

In 1995/6, using the minute Doppler effects on stars exerted by orbiting large planets, eight new planetary systems have been detected; examples are 51 Pegasi, 70 Virginis, and 47 Ursa Major. The nearset is Lalande 22185, a mere 8.36 light years away. In 1999 the star Upsilon Andromedae was found to possess a planetary system comprising no less than three planets, while in that same year the star Tau Bootis's planetary companion was observed in transit across the stellar disc by direct measurement of light intensity (photometry). The total count of extrasolar planets has now (Sep. 2000) reached over 40, and is rising rapidly.

It is hoped that the Hubble Space Telescope, outfitted with improved systems by astronauts in 1997, 1999 and 2001, or its successor, the planned Next Generation Space Telescope, will at last bring Jupiter-sized planets into direct view. Europe's Infra-red Space Observatory, launched in Sept.1995, is building on the work of I.R.A.S. Finding extrasolar planets makes picking up a needle in a haystack a menial task: since a planet might typically be 1 million times dimmer and several thousand times less massive than its parent star, separating out the feeble planetary signal from its neighbouring parent is going to be very difficult. For several years, it was thought that the Dutch astronomer Peter van der Kamp had detected a gravitational "wobble" in the movements of Barnard's Star during 25 years of painstaking photography-this has, however, not been substantiated. Somewhat extraordinarily, 2 extrasolar planets have been discovered as recently as 1994, and confirmed in 1995-orbiting-of all things-a neutron star. Any inhabitants will require a predilection for high energy Xrays.

Thus, it would appear to be well established in the Space Age, that we live in a dynamic cosmic ecology, in which planetary system formation is frequent, and that the biochemical "seeds of life" are widely scattered in interstellar space, needing only a fertile "womb planet" for maturation. Thus, Life was possibly generated on Earth from space-even if we fall short of accepting the more radical view of Sir Fred Hoyle that Life itself actually came from space.

Origins of Life

It would seem that Life emerged on Earth at an early stage. Fossils containing small bodies, believed to blue-green algae, have been found in South Africa and dated as 3,5 billion years old, while recently even more primitive microbacteria in Greenland rocks have been dated at 3,87 billion years. Planetary exploration has now shown us a solar system whose planets suffered continuing, heavy bombardment by leftover planetesimals until about 3,9 billion years ago. Earth's surface must have been intolerably hot and unstable-indeed, the "crust" must have been little more than molten basalt, and distinctly unfriendly to Life. Allowing time for conditions to settle down after this pummelling, and admitting that the earliest biochemical and primitive cellular evolution required some time, Life appeared very rapidly indeed on an evolutionary timescale. The transition from pre-life to Life seems to have taken place over a few tens of millions of years at most. The distinction between animate and inanimate is, at this level, a grey area, and defies exact definition. A virus, after all, is a crystalline chemical on a laboratory bench, but in a cell is a self-replicating invasive killer. I take as Life an entity that abstracts energy and raw materials from its environment to maintain its integrity, and which can reproduce itself from simpler chemical components according to a blueprint within self-replicating genetic material. On Earth, this means D.N.A or R.N.A, contained within a dynamic cell membrane. Both of these substances are assembled from four nucleic acids, strung out on a double spiral, whose backbone is made of chains of carbohydrate molecules and phosphate radicals. The information is carried in the sequence of nucleic acid bases-each set of three codes for the assembly of one of twenty essential amino acid into chains, which we call proteins. Of course, the code allows for full stops, switching on and off of genetic sequences, three dimensional configuration of protein structures, and more mysteriously, the "fourth dimension" in assembling a living organism-namely embryology. This development took place some time after the end of heavy bombardment, 4 billion years ago, and before the earliest known fossils, 3.5 billion years ago.

Another way of looking at this is that life evolved from relatively advanced precursor chemicals as soon as conditions were favourable. If Stanley Miller could advance from methane and ammonia to amino acids in just 7 days-and later experiments have taken these amino acids, and, using magnesium and clay catalysts, proceeded to simple protein chains in a few weeks-it seems reasonable that life could have evolved at an early stage. It is even more understandable when one realizes that cometary nuclei contain huge deposits of organic chemicals and water ice in deep freeze. However, on approaching the inner solar system comets heat up, with the spectacular results we have known throughout recorded history. Many of these would have dumped their cargoes on to the early Earth. Intriguingly, there is now increasing evidence for an outer collection of trillions of comets, left over from the primordial solar nebula at up to 1 light year from the Sun. This Öort's Cloud is believed to be a reservoir from which new comets are periodically dislodged into the inner solar system. It is now accepted that many stars have this envelope of primitive comets, and that there is some cross-fertilization between the Öort Clouds of neighbouring stars. If the interior of comets is a cornucopia of life precursors, it seems likely that Earth was simply seeded from beyond, and its origin on Earth was not such an uphill struggle after all. The possible early origin of life on Mars, and Europa, if confirmed, would strengthen the idea that Life's seeds originate in primordial propylids (accretion discs on their way to planetary system formation). Hence simple life is quite possibly widespread, requiring only good planetary soil for its germination. Intelligence, is of course another matter!

Simple life spread rapidly and easily in the warm waters of the early oceans, as continents formed, combined, and split up under plate tectonics-maybe twelve times before Life evolved beyond the single-celled stage. Single cells evolved into chlorophyll-bearing types, which tapped sunlight to convert atmospheric carbon dioxide into carbohydrates for energy, and into heterotrophs, which lived off the "proto-plants", or other, smaller organisms. The former were the cyanobacteria, and can be found today.

Slowly, the original atmosphere of carbon dioxide, methane, ammonia, and water vapour was converted to nitrogen, hydrogen, and-horror of horrors-oxygen. Later, some hydrogen drifted off onto space, while the oxygen created the first, and most serious pollution crisis in Earth's history. The oxygen, produced as waste by the early blue-green algae, built up over 1.5 to 2 billion years , until it reached toxic levels. New cells emerged which could survive in low oxygen atmospheres, but eventually, cells of a new type evolved, which could bind the fiercely corrosive and toxic oxygen, and convert it to harmless carbon dioxide, for release into the sea. By setting up a cascade of moderately energetic chemical reactions to "step down" the energy of oxidation, the new cells could cope, and being opportunistic, Life soon managed to take advantage of the new situation. The controlled stepwise oxidation, and the ability to store the energy as triple phosphate chemical bonds, made available to the heroic survivors of the oxygen crisis 38 times as much energy as mere photosynthesis. This took place between 2.3 and 1.8 billion years ago. Sophisticated compound cells, which consisted of the symbiosis of four different microbial species, led to more active, self propelled cells, which soon differentiated into many different species.

But then, about 1 billion years ago, Life evolved into colonies of single cells, or organisms. At first, each member of the colonies was very similar to each other, but in time they came to specialize. The causes of this change are lost , but the first multicelled organisms seem to have arisen quite rapidly. Sexual reproduction also appeared. This allowed for much greater rapidity of genetic change, since the random mixture of genes from two parents, rather than mere replication , allowed for new results, upon which natural selection could exert new pressures. Evolution could thus accelerate; indeed, the main invertebrate groups appeared in an astonishingly rapid period known as the Cambrian Revolution, lasting a mere 7 million years.

At this point it is worth raising a philosophical question. Reductionists say that all evolution is without purpose, or that, at best, D.N.A acts only to replicate itself, so that all species are solely vehicles for transmission of "selfish" D.N.A. They have to ask themselves : "Why, after nearly 3 billion years of completely successful and undisturbed occupation of the oceans by simple, eternally self-replicating, single-celled organisms, did Life suddenly evolve into the vastly more risky business of multicellular existence?" An existence, moreover, which is subject to death, and which must find a mate for reproduction, on top of all life's other hazards. In addition to all this, metazoans are much more tied to a local ecology than a blue-green alga, which can live off sunlight, a few simple chemicals, and can reproduce itself as long as there is a reasonable amount of sea-water about at the right temperature.

Life invades the Land

Still more does this question arise in considering the next major development-the conquest of the Land. This enormous step was taken about 400 million years ago, at first by plants, and later by ancestors of the modern lung fishes. These creatures, even today, are often beached on the shores of tidal waterways. Initially, the ability to breathe perhaps arose as an insurance against drought.

The settlement of the Land required the loss of many advantages of marine life. In the primaeval seas , organisms were weightless, supported by the warm Ocean. Oxygen diffused into open gills, and food was easy to find. Maintaining water balance was easy-there could be no dehydration or desiccation in the sea. Yet at this time, organisms forewent all these advantages in favour of a new environment. On dry land, there was gravity and solar ultraviolet radiation-an earthworm will perish within an hour on exposure to direct sunlight. Creatures could no longer float or paddle about, but had to invest enormous amounts of biological capital into fighting gravity-only now that we experiment with "zero" gravity can we appreciate the true cost of conquering dry land. The new land animals had to develop much more powerful hearts and circulatory systems to pump oxygen bearing blood against gravity, as well as more active systems for extracting oxygen from the air. Fluid balance became a new demand; the continual availability of water in the sea could no longer be relied upon. The skin became a flexible, tough, and efficient guardian of the primaeval interior: the fact that blood is 0.9% salt attests to the reality that we carry within our veins the oldest living fossil. For our blood retains preserved, in many details, the waters of the ancient Ocean. The skin became the interface between the familiar and the harsh new atmosphere of dry land, conserving chemical and liquid composition, and protecting against radiation, wind and storm, and extremes of temperature-there were Ice Ages in Permian times, 220 million years ago. In that early conquest of an entirely new element, the skin was as essential to the safety of Life as the new skin we must take into the beckoning new element of today: the skin was Nature's first space-suit.

Looking back with 300-400 million years of animal occupation of dry land, what prizes justified this vast and hazardous change? From a reductionist view, the whole enterprise, like the development of metazoa, and sexual reproduction before it, was unnecessary. The oceans comprise about 2/3 of the available living space of the planet, so that, from the point of view of Life as a whole, the territorial gains were not impressive. However, in compensation for the burdens of gravity, radiation, temperature extremes, and fluid balance problems, there were several clear gains.

Firstly, the oceans contained a dissolved oxygen content of 4%, whereas the atmosphere contains, at present, 21% oxygen, and, even in those days, much more than the niggardly marine 4%. This meant that there was a far richer energy source available than anything as yet seen, and metabolism could proceed at a far higher level.

Secondly, a great diversification of habitats and species became possible: there are greater contrasts between mountains, valleys, deserts, tundras, temperate and tropical zones than between the more homogeneous oceans. On top of this, the continental "dice", and attendant climates, are shaken up relatively speedily by continental drift. The collision between the Indian subcontinent and Asia, for example, has thrown up the Himalayas in 50 million years, while, more recently, the Pacific Archipelagos have appeared in the last 5 million years. If one conceives of Evolution as a creative force seeking more diverse and sophisticated organisms, the ability to change, local variations, and exposure to cosmic radiation make for a more rapid pace of evolution on land than on sea-and so, indeed, it has proved.

Thirdly, life on land allowed the development of new senses and abilities-long range sight with binocular vision, the ability to manipulate and "work" the environment, the power of flight, and the ability to discover, and tame fire. These would have been impossible to life in the sea, and are vital for a technological civilization.

It will be objected that the Cetaceans-whales and dolphins-constitute a communicating, intelligent species, maybe with a culture all their own. Quite so, but it is of a different, non-manipulative nature, and, in any case, the Cetaceans are mammals which developed on dry land 60 million years age, and returned to the sea. The "jury is out" on the question of the whales and their development of Mind-but, as will be seen later, it is the development of a technological civilization which, in purely utilitarian, Darwinian terms, constitutes the flower of the vast enterprise of Life on land.

Towards Mind and Purpose

What of the development of Mind? Is this a random event without meaning, and, from a reductionist view entirely unnecessary-

or is the development of consciousness a prime goal of a Creative Evolution? Is it simply a development from a peculiar situation arising in East Africa 6-8 million years ago? One might have thought so, but recent discoveries relating to the times of the dinosaurs tell a startlingly different story. Much of this material has only become known in the last 20-25 years, and proof of the general scenario to be outlined is only a few years old-details are still coming in. The full implications are thus unfamiliar to many people, and will probably not be fully accepted for a generation-but they are inescapable.

Everybody knows of the dinosaurs as gigantic behemoths weighing several tons, lurching about like dumb colossi: it is now realized that these giants were only the most visible members of a whole, complex range of creatures ranging from the size of a chicken to the incredible Ultrasaurus. To see all dinosaurs as monsters would be like seeing all mammals as rhinos or tigers. Then there are the anomalies. For a start, there is evidence from microscopy of bones that some dinosaurs were not cold blooded , but warm blooded, like birds. The predator/prey ratio in dinosaurs supports carnivorous dinosaurs of high metabolism, while the length of stride of predator's foot prints suggests agile fast movers, not lumberers. Towards the end of the Cretaceous Age, 75-65 million years ago, there were the direct ancestors of the birds, the formidable Deinonychi, who walked-and ran-on two legs, and developed light balancing forelimbs. Within this general group were some remarkable species-little more than 2 metres tall, which had prehensile fingers, as well as a pair of eyes situated on the front of their heads, rather than on the sides, as in the classic reptile. Such creatures hunted in packs, and were undoubtedly skilled and clever hunters. With binocular vision and prehensile hands, with the beginnings of opposable thumbs, added to warm, mammalian style metabolism, we have a creature that was as much a precursor of Intelligence as another small contemporary, the Tarsier, whose descendants now survive in the forests of Madagascar. These Tarsier ancestors are now generally recognized as the progenitors of the order of Primates. This was well described in a BBC TV "Horizon" programme, dealing with the extinction of the dinosaurs. Canadian palaeontologist Dale Russell has suggested, entirely reasonably, in view of the eventual meteoric rise of the Primates, that, without the sudden demise of the Dinosaurs, the Raptor/Deinonychus group was well down the road to Intelligence. Using some legitimate extrapolation, Russell drew an image of today's would-be "Dinosaur Man"-a somewhat chilling sibling of Dan Dare's infamous enemy, the Mekon! Given 65 million years, the "Mekons" would by now be the true masters of the Galaxy. Why, then, is that not the current state of affairs, and how has the lowly Tarsier come to have descendants standing on the borders of space?

The riddle has arisen because the records show that this whole vast, flourishing world system of highly diverse and capable animals, together with 85% of all species then living vanished in the twinkling of an eye, geologically speaking, after flourishing for well over 100 million years. In the "watershed" layer between the dinosaurs' Cretaceous Era, and the succeeding mammalian Palaeocene Era, is found, clearly, and around the world, a 20cms. thick layer of deposit with 30-50 times the normal concentration of iridium, an extremely rare silvery metal. This metal is usually found only in trace amounts, in association with other equally rare metals, in deep rock deposits. While occasionally thrown up by volcanic action, it is never found on its own, in such anomalous concentrations, in sites all over the world, on all 6 continents: immediately above the Iridium layer is a layer of soot, as from a universal conflagration. The late Professor Luis Alvarez in 1980 put forward the thesis that this was deposited by collision with an asteroid 10 kilometres in diameter, bringing with it primordial elements, including iridium, left over from the birth of the solar system. Alvarez proposed that, 65 million years ago, a 10 kilometre sized asteroid hit the Earth at about 60 kilometres per second, throwing up vast quantities of dust, steam, and debris into the upper atmosphere, leading to total darkness for several years, a marked chilling of the climate, and loss of the lush vegetation which had sustained the dinosaurs. All large land animals, and all life except for a few plants and those small animals which could live off carrion, perished.

What evidence is there ? Firstly, there is the sharp break in the fossil layer at the right time, 65 million years ago, with a band of several thousand years thick in which no large animals are found.

Secondly, there is the anomalous quantity of iridium found in the boundary layer, and its global distribution.

Thirdly, there is evidence of other "mass extinctions", some with their own iridium layers, over the last 600 million years. Some half a dozen have now been identified, at intervals of approximately 100 million years or so.

Fourthly, it has been found that several small asteroids, which have similar lethal potential, are of the right order of magnitude, and do approach the Earth-the Apollo, Ares, and Aten classes are cases. Hermes approached within 640,000 kilometres of Earth in 1937, while Cold War nuclear monitoring satellites picked up kiloton sized explosions in the upper atmosphere of Earth-grazing debris several times per year. In 1989 and May 1996, 300 metre asteroids passed close by the Earth at less than 400,000 kilometres with only 2 weeks' warning.

Fifthly, in the early 1990's, a clear asteroid impact crater was located straddling the coast of Mexico, off the Yucatan Peninsula. This has been dated as 65 million years old, and the result of a collision at speed with a 10 kilometre asteroid. Knowing the approximate number of asteroids in Earth crossing orbits, with their sizes, and orbital characteristics, it has been possible to compute the likely frequency of collision between the Earth and any given size of space debris. For a body of 10 kilometres in diameter, the expected frequency comes out at one in 100 million years- which is what is observed in the fossil record. More recently, it looks as if there were a number of collisions at the time of the dinosaurs' extinction, which suggests that Earth passed through a collection of objects.

Sixthly, over 130 impact craters more than 100 kilometres in diameter have occurred within the past 2 billion years, and have been detected from the Space Shuttle. Since our planet is two thirds covered by water, the true number of large impacts must be at least 400 over the same period, while, allowing for plate tectonics and weathering, the number is likely to be yet larger.

Finally, we have seen, in July 1994, the impact of Comet Shoemaker-Levy 9 on Jupiter. No one can seriously doubt that such a disaster could hit the Earth, and that, in that event, the results for any human civilization based on Earth, and indeed, of all "higher" lifeforms would be terminal.

There is one more point to be made about Shoemaker-Levy 9; it was only discovered 2 years before its onslaught on Jupiter, after it had been altered in its orbit and torn into 21 fragments. Thus at any time we could be given 2 years "notice to quit" the planet by a sharp-eyed astronomer with a good calculator, and, if our civilization is confined to Earth at that time, all we could do is pray, with a possible slim hope from an untested and probably hastily improvised nuclear shield.

A startling new picture emerges. 70 million years ago, intelligence was, far from being a wildly improbable freak, advancing along three fronts towards three intelligent species in embryo-"Dinosaur Man", Mammalian Primate Man, and the Cetacea, and that all three were advancing towards consciousness before disaster struck in the shape of a large cosmic rock. The main goal of Creative Evolution is not mere D.N.A survival and replication-after all, that limited goal could survive even a succession of major collisions-but intelligence, and, in particular, Mind. Since the debacle of 65 million years ago Evolution has been working , using the available surviving material, to produce a species capable of doing the one thing that can ensure the survival and growth of Mind, as opposed to mere genes-namely, dispersal from this one planet before the laws of celestial mechanics interfere again, and thwart the whole enterprise.

The emergence of curiosity, aggression, and the desire to manipulate, understand and eventually dominate the environment can all now be seen as pre-requisites for fulfilling the prime imperative of Evolution: to reach consciousness and intelligence to ensure not only survival but also advancement in the real, final home of Mind-the great ocean of space itself. Thus the answer to the question of what distinguishes Humanity from all other species, giving us meaning and purpose, is not metaphysical, or subject to sectarian ambiguities-it is practical, and immensely motivating. It is simply that, alone among 500 million species, we have gained the ability to leave it, and to continue the great adventure of Mind on a larger canvas. In conclusion, then, the conquest of the land is a stepping stone for Life traversing from one ocean to the next infinite one, with the entire 300-400 million year sojourn on land as but a stepping stone in an epic, weightless journey.

Chapter 2

Mythology and Religion-"The Fires of Prometheus"

In the last chapter we explored the origins of Life, and of the world we live in. I put forward the idea that evolution is a creative force, with a sense of direction, and that this direction leads , via the growth of mind, technology and intelligence, to the conquest of space.

For the sake of survival, let alone betterment of the species, the extension of life into space is now clearly a necessity, considering the climatic, ecological, and cosmic disasters on our planet. Earth is not an island spaceship , but part of a Universal ecology.

If there really is a plan leading towards a future in space, then one would expect to find a leaning towards the sky and space as major themes of human psychology mythology, literature, philosophy and ideology. Carl Gustav Jung was a very widely travelled observer of subconscious psychology, and made the break with his mentor, Sigmund Freud, over the question of whether all or most dreams and psychological disorders take their origin from the repression of sexual desires, specifically the taboo ones of incest. Jung refused to reduce all subconscious motivations to the sexual. From travelling the world and meeting people from many cultures, he derived the ideas of "Archetypes" and the "Universal Unconscious".

He discovered that familiar themes occur in dreaming among peoples of all races, sexes, and creeds, from the Kalahari Bushmen to angst-ridden central European sophisticates, often involving the figure of a "Wise Old Man" , or the lotus flower images of flight, or of the "Earth Mother". These symbols he called "Archetypes", and he considered the to be manifestations of something deeper than the individual personality. They also appear as recurring themes in literature and the arts, giving them universal appeal. The wizard Merlin, Star Wars' Obi-Wan Kenobi, and Gandalf are really all the "Wise Old Man", helping youth on its quest. He considered the Archetypes as the signature of a universal mind. The racial mind exists at a deeper level than the Freudian Id and Ego, and has a long history in human philosophy and religion. The idea of achieving fulfilment through the submergence of self in the greater or universal Self is at the root of many religions, most markedly in Buddhism, where the God is seen as an impersonal, unnameable "suchness": this is taken to great extremes where the idea of hurting another creature is deeply repugnant because, in part, it involves hurting an aspect of one's Self. In the Jungian scheme of things, the myths of humanity were not mere fireside entertainment , but drew their power from the fact that they strike a chord in human consciousness, in the racial experience of our species. The Greek myths of Oedipus and the Judgement of Paris still retain an emotional power in the 20th century minds, despite vast cultural differences from their intended audiences. The idea of a man who seeks to avert his fate, by separation from it, only to lead himself inexorably into the mire, is a story of all ages, and the blind prophet Teiresias who reveals the truth and foretells the inevitable doom is, again, the archetypal Wise Old Man, while the daughter Antigone, as Oedipus' sole companion in exile, like Lear's Cordelia, is another archetype. Indeed, according to Jung, the myths of Mankind were the means by which the collective unconscious communicates with conscious minds; the process continues in our modern world, using the old archetypes in modern dress.

Early Cosmic Connections

If Humanity is truly destined for the stars then mythology, past and present, ought to reveal an interest in that direction, and, indeed, they do. The first races of modern humanity, Magdalenian, Aurignacian and Cro-Magnon, preceded the end of the last Ice-Age, about 12,000 years ago, and left no written records. However, they painted caves with contemporary hunting scenes, and wore clothes of considerable craft, ivory and precious stones. In some Magdalenian caves there have been found collections of neat round pebbles, evidently used as counters of some kind, on which were depicted images of the Sun, phases of the Moon, and caricatured constellations; this is not put forward some new "ancient astronaut" theory, but merely to suggest that the earliest known symbols to be drawn by humans relate to the subject of this book. The phases of the Moon were thought worthy of observation and recording long before they could have any conceivable meaning.

The observations of Sun, Moon, and constellations were calendrical in origin. The planting and reaping of crops by settled communities, and, before then, the advent of spring for migration purposes necessitated study of astronomy, and constituted the first exploitation of space resources!

In Egypt, the flooding of the Nile, vital as it was for the inundation of the fertile Delta region , was marked by the appearance of Sirius or Sothis in the summer sky, and led to the first calendar. The inadequacy of this calendar led progressively to our Gregorian one of 365.2422 days to the year. Recent discoveries of the alignments and structure of the Pyramids of Egypt, have led to a remarkable recent publication-"The Orion Mystery"-by Robert Bauval, in which it is suggested that the Pharaoh, on demise, resumed his identity as the god Osiris, identified with the giant hunter-figure of Sahu, or Orion, alongside the heavenly Nile, known to us as the Milky Way. Thus the prayer "as it is in Earth, so it is in Heaven" held for the Egyptians an older, and far more literal meaning. The Egyptian monarchy, its funeral and coronation rites, were thus literally dependent upon the correct reading of the pageantry of the night sky, and the function of the Pyramids was to integrate the world of Man with the Kingdom of God in the person of the Pharaoh. There is much more of interest in this work, and remarkable discoveries about the religious life of ancient Egypt are likely-but enough has been said, I think, to justify my belief that the science of astronomy has played a vital role in human civilization from the outset. The Egyptian myths, on another level, of daily death, rebirth of the Sun, and the annual death and rebirth of Sirius are clear attempts to describe and account for the most vital and obvious cycles of the day, and the seasons, while the lengthening of nights over the year was colourfully explained by the myth of Hades and Persephone in Greece. These calendrical myths are early uses of the resources of space, and without them no systematic agriculture or government could ever have been developed. Long before people thought of venturing into the sky, they had "conquered" Time, and learnt how to measure it and predict its passage. From this came the ability to plan and foresee, with all their implications. For some, no doubt, it meant a loss of "innocence" or "freedom from Care for the morrow", as implied in the myth of the Fall of Man from a state of grace, for, ever after harnessing of Time, people have been yoked to the tyranny of the clock. One measure of the complexity of a civilization must surely be the extent and accuracy whereby it measures and records the passage of Time.

Early people also observed the constellations in the sky, and being by nature inclined to impose patterns in their surroundings, saw in them people and beasts-larger and more perfect replicas of the animals they shared their world with. In many cases, however, they led on to legends which reflected their deepest yearnings. For example, the Lady Andromeda chained to a rock and rescued by the valiant Perseus is a drama re-enacted every night in the winter skies of the Northern Hemisphere, and reflects the age-old saga of the Hero, who rescues the Maiden after countless wild adventures, with the help of the gods from heaven and their magical devices and spells-a metaphor for the rewards offered to the Adventurer and Voyager. The fierce Lion, subdued by Orion, commemorates the subjugation of the strongest of the beasts by human wit and weapon-an early triumph of technology.

The debt owed by humanity to the Sky, and its role in launching humankind on the road to civilization, and to the stars, is further shown in two considerations. Firstly, long before smiths ever mined and smelted iron ore, this most valuable of metals was freely, if spasmodically, available in the shape of meteoritic iron ; there seems little doubt that the Ancients knew full well of its origin in the sky, and drew from it proof that the gods lived in a real Kingdom.

Initially, the Swastika is believed to represent the rebirth of the hibernating bear in the spring, as it rises towards the zenith in the summer, and the rebirth of the natural cycle, but later the triumph of light over Darkness as in Manicheism, and, perhaps, coincides with the Cross symbol of the resurrection bringing the Christian triumph of Good over Evil, and Life over Death. This is not meant to be a comment or "explanation" of the divinity or otherwise of Christ and his mission, nor to demystify it, but to add another dimension, or resonance, to the symbolic resurrection of the Cross. The Cross, in other words, is an archetype, and, as such, I believe, it may well be derived from the Heavens.

The ancient Egyptians also venerated a "Cross of Life", and rebirth called the Ankh. This is frequently worn as a talisman of costume jewellery to this day, while to the Buddhists this symbol appears as the Wheel of Life and Death. Others have cited the swastika symbol as evidence of a widely observed Great Comet.

The foregoing, I believe, illustrates the importance of the sky and stars in forming myths and symbols, and the way in which people attach great importance, throughout the ages to these symbols, formed as they are in their innermost being. It argues strongly, I feel, for a central role for space in Humanity's deepest imagination, and our intuitive awareness of the links between our race and the heavens.

Wisdom from Heaven

The linkage found further expression in many of our most famous myths; the Greeks and the Babylonians believed that civilization in the shape of fire, metallurgy, laws, architecture, and medicine were gifts from heaven. The Greeks told that Zeus, father of the gods, achieved supremacy only after conquering his father, Saturn, who was bent on eating all his children; after a terrible struggle in which he was aided by the giant Titans, he succeeded in vanquishing Saturn, and enthroned himself as father of the gods. Zeus called together the gods and Titans, and bade them create Man, who would cultivate the Earth, and honour them with sacrifices. Man was to stay servile, and there was an absolute prohibition on their acquiring Fire, as this would make him too much like the gods. However, one of the sons of the Titan Japetus, Prometheus, loved Mankind and secretly brought down the gift of fire from heaven, with which our ancestors proceeded to forge metals and build civilization. It is likely that our first use of fire, as well as iron, was derived from space, as blazing meteorites or thunderbolts. Eventually, Zeus, seeing the smoke of many fires, caught Prometheus and punished him by leaving him chained to a rock. By day a great vulture was to eat out Prometheus' liver and, by night, left the liver to regrow for a further meal. Man was punished by being given the gift of Pandora's box, from which escaped all the ills that now afflict Mankind. The Greeks saw Zeus and the gods, not as all-powerful, loving Creators and protectors, but as projections of natural forces and that the correct mode of life was to live in balance with them. The morality was not such an absolute one as ours, with Man bound to obey a Good God, but based on Man in balance between unpredictable forces.

It is not therefore a blasphemy to suggest that, in reaching for the stars, we have tamed the gift of Prometheus, and are moving up to take our rightful place alongside Zeus at Mount Olympus, as Prometheus the benefactor would surely have wished. The true importance of the legend of Prometheus for our story lies in two natural ideas, implied in the tale, and in many of the creeds and philosophies which have found acceptance among Humankind. They are, simply, these;

Firstly, that Man is not a perfect creature; there is room for improvement, and the possibility of achieving it, and, secondly, that the source of Wisdom and Power, and the environment necessary for its accomplishment, is to be found in the sky. In the Prometheus story, we see both these ideas at work; Fire, the means of improvement in our lot, has to be brought down from Heaven, by an agent working for our interests against the gods.

In ancient Mesopotamia, the shadowy Sumerians emerged in the land bordering the Persian Gulf between the Tigris and Euphrates during the fourth millennium B.C. They are believed to have been immigrants into the region, perhaps from Central Asia or India. In any event, they acquired civilization very rapidly; by about 3000 B.C., they were farming systematically, working in bronze, writing, and building vast "stairways to the gods". The Ziggurat of Ur, and its near contemporary, the step pyramid of Saqqara, were triumphant monuments to the newly acquired arts of civilization, and the genius of the rulers who could conceive of such new and unearthly shapes. They also produced the first known written work of literature-the Epic of Gilgamesh. Its date of writing is unknown, but is believed to be 1500 B.C., and is probably derived from far older sources. It tells the tale of a hero-prince Gilgamesh, who seeks immortality, and freedom from ill. To this end, he rides into Heaven to seek out Enkidu, King of the gods, to cajole him to yield the secret. He sets off on the back of a gigantic eagle, at such a speed that he feels himself an increased weight being pulled back towards the Earth and, looking back, sees a patchwork quilt of green fields and blue rivers and seas from his vantage point in the sky.

Ancient Astronauts?

The point is not that this is proof of the visitation of ancient astronauts, although it would be foolish to discount this entirely; von Däniken might have the last laugh on us all. Rather, I see this as another illustration of the idea that people have always sought betterment from the skies. In describing the medical aspects of space development later, we shall see that Gilgamesh was not, actually, very far in error.

Later in time is the "Berossus fragment", a history of Babylon and the preceding Chaldeans and Sumerians by a high priest named Berossus, writing in about 400 B.C. This fragment survived into Hellenistic times and was transmitted to us by the Greek historian Plutarch. Berossus gives us the remarkable accomplishments of the Sumerian people. In earliest times, he says, the people of the area were nomads and illiterate barbarians, when a race of amphibious sky people called the Annedoti (unsightly ones) appeared from the waters of the Persian Gulf by day, and had to return to the water by night. They came from a watery world among the stars, and were not gods, although endowed with great intelligence. For many years, they ruled over the primitive people, teaching them the rudiments of civilization, laws, architecture, metallurgy, writing, mathematics, and astronomy, and then returned whence they came. This story, wrote Carl Sagan in "Intelligent Life in the Universe", is probably the closest thing we shall ever find to a true "contact myth", fulfilling many criteria laid down as possible evidence of visitation from another world in the past. These are, firstly, a story of visitation in which the visitors are clearly stated to be non-divine; secondly, a description of the visitors in terms not so vague as to be meaningless or biologically absurd; thirdly, evidence of actual transfer of information with or without artefacts, fourthly, a plausible timescale for the interaction with humans-over several decades or centuries-which does not violate the laws of astronomy or physics, and, fifthly, evidence of possible similar contact elsewhere on the Earth with similar results.

As Carl Sagan pointed out, there are no artefacts, and no independent evidence of other contacts, unless one accepts that the Dogon people of West Africa worship the Dogstar Sirius, and its unseen companion, Sirius B, as a result of a similar contact. The case for genuine contact remains tantalizing, but much more will be needed to quell the sceptics-and rightly so, considering the magnitude of the proposition! However, my proposition remains that both the Babylonians, and the Dogons-and by extension the Ancient Egyptians-looked to the sky and its possible denizens as the source of wisdom and power, rather than, say, the depths of the sea, or even their own Inner Consciousness. All of these latter are much more easily accessible than the distant sky; the Ancients could dig, explore caves, lower ropes into the sea, or even use primitive diving bells (Alexander the Great). They could also indulge in mysticism, transcendental meditation with or without the dubious benefit of drugs-indeed, it could be said that the Ancients were more in touch with their own Inner Selves than our sophisticated, harassed, "performance-related" modern business people and executives. There was, however, no way they could realistically expect to attain wisdom, or anything else, from the utterly unattainable Sky-yet it was there, over and over again, throughout the ages-that people have, searched for the greatest goals of Humanity. This idea stands, irrespective of whether we were visited by ancient space travellers. We were launched towards a technological civilization in a metaphysical way from the stars, and the end result is the same. The desire to go up into the sky was there both in the Epic of Gilgamesh and the Greek legend of Icarus and Daedalus. These sought to fly from Crete using a machine made from bird's feathers stuck onto a frame with wax. The impetuous Icarus flew too high, and suffered a catastrophic malfunction of his aerial surfaces, perishing upon re-entry, while the wiser Daedalus made a soft landing! The idea of flight, both in the air and space, was therefore not mere metaphysics, but one to be pursued rationally, with the most advanced knowledge of the times. It has been this way ever since.

The dreams of flight and space travel also appear in the legends of ancient India and China. For China, it was flying sky-dragons which gave birth to the Middle Kingdom while as a side-swipe at those outside the Empire, releasing dung into the Pacific from which sprang Japan. In the epics Ramayana and Mahabharata, Rama has to fly from India to Lanka to rescue his bride Sita; the flight takes about 40 minutes in the epic, by flying chariot. This is about the same as by Air India today! Also, there are long and rather detailed chapters on the construction of flying chariots used in aerial and celestial warfare and transport in these epics. The possibility of travel to other worlds crops up here, and also in the Rig Veda or Book of Knowledge, in which the Earth is described as being composed of three spheres of different sizes, one within the other, while the Moon is covered with particles of glistening glass balls. I take the finding of glass spheroids by the Apollo astronauts as an interesting coincidence, but surely the unknown author(s) of the Rig Veda deserve credit for being the first to conceive of the Moon as a world, with a composition, which might be reached, and utilized. It is also believed that the flying carpets of Haroun al-Rashid's Arabian Nights derive from some of the legends of India narrated to the Grand Caliphs' enterprising traders.

"Heaven on Earth..."

With the Ancient Egyptians, the credit for founding civilization also goes to the sky, in the person of Thoth, called by the alchemists Hermes Trimegistos.

He earned this title by bringing the gifts of Law, Medicine, and Architecture. In legend he descended from Heaven on winged feet and ruled over the early Egyptians. He fathered a son, Aesculapius, by a mortal woman, who began the practice of medicine. Aesculapius carried a magic wand around which were wound two serpents in a double helix, the symbols of royal power. This stick, the Caduceus, is even now the internationally recognized symbol of the physician, the sceptre of Aesculapius. Aesculapius was the Greek name of a real man, whom the Egyptians called Imhotep. This man must have been one of the most influential geniuses to have lived, although little known. He lived early in the 3rd millennium B.C., shortly after the first Pharaoh united Upper and Lower Egypt, and was an architect, astronomer, high priest, doctor and lawgiver. One of his titles was "Chief of the Observers"-an indication of the importance attached to astronomy/astrology in those days. It would most likely to have been his responsibility to predict the flooding of the Nile, and to give the go-ahead for planting and reaping of the vital food supplies. During the Fourth dynasty, he and his master, King Djoser, embarked on a gigantic, hitherto inconceivable, enterprise. They were to build a "stairway to Heaven", for the Pharaoh; as the embodiment of Osiris on Earth, his resurrection and ascent to the Heavenly Egypt in the Sky after his decease on Earth was essential for the well-being of his subjects, as well as for an orderly succession. On this principle depended the order of the newly created, unified Empire of Egypt, with its vast nationwide irrigation projects, granaries, and ordered life. To ensure this prosperity, the body of the god-king had to be preserved as an enduring symbol of the New Order, and for the stairway to Heaven represented by the ascent of Pharaoh's spirit to the realms of Osiris, Imhotep proposed a revolutionary new scheme; the "tomb" was to be a pyramid executed in stone. The result was the Step Pyramid of Saqqara, which endures to this day. Robert Bauval tells us, in the intriguing "Orion Mystery", that the pyramid complex, begun at Saqqara, was part of a far more ambitious scheme-in the astronomical alignments of the greatest of all the pyramids, there was to be embodied the rising of Orion's belt, to its maximal height above the horizon, at the epoch 2,450B.C, revealing a knowledge of equinoctial precession over a very long time-and, an eventual recreation of the "Heavenly Kingdom" on Earth, in the form of a replication of Orion's belt. It is possible that the assignment of a divine father, Thoth, upon Imhotep/Aesculapius, represents a memory of the original discoveries in observational astronomy which led Imhotep and Djoser to launch what was to become, over the next few centuries, the largest building project in history, or that Thoth was actually a living astronomer, or founder of the guild in which Imhotep and others trained. We shall never know, unless Imhotep's tomb is one day found.

One must imagine the scepticism which greeted this enormous, megalomaniac project. The materials were untried, the forces involved unknown, the labour and expense colossal. To Egyptians in 2600 B.C. this proposal must have seemed more far-fetched and grandiose than our ventures into space seem to us, and yet the will triumphed and the foundations were laid for a civilization which was to endure for 2500 years, and to capture the imaginations of people on Earth for as long as the capacity for wonder and admiration exist. If Bauval's ideas are even partly right-and several scholars are now considering this-the whole Pyramid project attests to a culture that could take a very long-term view ; how many of our ephemeral politicians could conceive of pushing through a project which would not be finished by their great-grand children? It is also apparent that, in their belief that the pyramids were essential "launchpads" for the Pharaoh/Osiris to reach the Kingdom-literally-in the Sky, at a specific location, and that this was vital for the good of Egypt-Imhotep and Djoser were, *mutatis mutandis*, precursors of our own space activists and pioneers. The idea of space travel, and the improvement of the Human future by moving into space, is age-old and instinctive to Humankind-rests!!

Here we have, *par excellence,* the idea of the sky and space as formative for human development, and for our future. This idea continues right through to modern times, with popular, although, superficial interest in astrology, as well as the astonishing popularity of the "Ancient Astronaut" books exemplified by Erich von Däniken. With remarkable timing, in 1968, Erich von Däniken, a Swiss hotelier, launched upon the world a book called "Chariots of the Gods", which takes as its central idea the visitation to Earth by superintelligent beings from another planet, who created Man out of apestock, and played a part in founding our greatest civilizations, chiefly by building artefacts thought to be beyond the capability of people at the time. Much to the amazement of most people, including von Däniken himself, this book and its sequels sold in the tens of millions, and found many willing to listen. It is true that many of the "impossible" facts have been explained in terms of human ingenuity, or have been proved inaccurate; I can certainly testify that his astronomy is frequently in error. To cite an obvious example, he points to the Pleiades as one point of origin for the gods from an Red Indian Legend. All astronomers agree that the Pleiades, or Seven Sisters, are a young cluster of stars, less than 60 million years old, and therefore highly unlikely as the breeding ground of a mature, space-faring civilization. His observations on the subject of Easter Island are now shown by archaeologists and palaeobotanists to have quite another and much more disturbingly relevant explanation than that of ancient astronauts. Indeed, the "Riddle of Easter Island", far from demonstrating the reality of past alien astronauts, proves the vital necessity of contemporary human ones! However, it is not my purpose here to endorse or refute his thesis: others have performed this task more ably than I could do, and, in any event, he may yet have the last laugh. It is hard to prove a negative.

The point is that the ancient mythology I have described, plus the modern legends propounded by von Däniken, as well as the whole U.F.O phenomenon, point to a very definite human interest in the whole subject of the Heavens, and the desirability of our exploring, and settling there. There is a general subliminal perception that an important part of what we are as human beings comes from the sky, and that is there that we must look for our future. So much is this so that this skyward inclination of humanity can almost be taken as a hallmark of true civilization, and that its absence spells barbarism.

To return to Jung; he lived to see the rise of the U.F.O phenomenon, and saw in this an archetype of the search for meaning, without which the human psyche is incomplete. The U.F.O crews are almost invariably benign and omniscient, and are easily recognizable as the latest manifestation of the "Wise Old Man" archetype.

Life's Unfinished Symphony

In the light of the evolutionary course of humanity, propelling life onwards, and outwards into space, and the strong mythological and psychological attractions of the concept to so many millions over the Ages, it seems increasingly plausible to link the two phenomena as being manifestations of our true destiny-the habitation of space. This is in no way denied by the fact that, in particular periods, human civilizations decay, and the instinctive drive falters for a while. Change occurs in fits and starts, rather than smoothly.

We have looked at the idea that Humanity has, for many ages, imagined wisdom, civilization and technology as a gift to be won in the heavens; let us now consider another equally powerful idea-the incompleteness and possible further development of humankind.

The idea of Evolution, or the replacement of "lower" species by higher ones with the passage of time, is found in many ancient writings. Heraclitus describes life as beginning in the water, as fishes, before reaching the stage of frogs and lizards, then horses and dogs, and next, monkeys and people. Similarly, in the Rig-Veda, such successive stages of living things are described with the idea that "Krishna Man" or perfect Man was to follow us. The Buddha also taught that humanity progressed from life to life by re-incarnation until rising above earthly desires , we become Buddha or Enlightened, one with all things, not separate, yet not extinct.

However, it does illustrate the belief that, in our present state, we are not complete, and are still riven by conflicts between lower selfish animal yearnings, and the more noble aspirations-and that, until we overcome this conflict of desires, we are imperfect.

In essence all the religions widely practiced at present-Judaism, Christianity, Islam, Buddhism, the Bahá'í Faith, and, in its own way, Marxism have in common the belief that humanity has not yet attained a perfect state. They are speaking mainly of the spirit-except Marxism!-but the idea is the same. The philosopher Plato realized similarly that Man was not a complete form-he describes the human personality as resembling a chariot driven by the two horses of Emotion and Intellect, with the driver, or Mind/Conscience, in control, ideally, pulling in the two unruly horses. In the perfect, or complete, person, all three would naturally pull together without coercion. If this situation did occur, it would be argued that there would be no morality, since, if people did naturally that which was highest and best without coercion, conduct would not be good in avoiding prohibitions, or bad, in avoiding virtues-but simply natural and rational.

This may be the true meaning of Nietzsche's philosophical work "Beyond Good and Evil", and is, perhaps, a realistic reinterpretation of the Nietzschean "Superman". Nietzsche maintained that the challenges and conflicts of evolution would eventfully destroy conventional morality with its small-minded hypocrisies, and replace it with a natural morality in a world of supermen-an idea which has been greatly muddled and abused, partly by Nietzsche himself in his later, madder, years, but more spectacularly by men who found that his phraseology suited their far from superb actions. Nietzsche himself had no time for anti-Semitism, the overweaning state, or the spiritually demeaning ideas of Nationalism-points which his Nazi worshippers conveniently "forgot".

A good and striking modern film showing the evolution of Humanity and the conquest of space as a myth for the 21st Century is "2001 :a Space odyssey". It is no surprise that Strauss's "Also Sprach Zarathustra", based on Nietzsche's work of that name, was used as the opening theme music. For in one 15 second film sequence the essentials of what Nietzsche was aiming at were well shown. In this sequence an early man, having used an antelope's thigh bone as a weapon for the first time, hurls it high into the air in celebration; in vanishing from sight, the bone travels 300 kilometres and 3 million years, and becomes metamorphosed into a space station.

This is a most powerful visual demonstration of the idea I outline in this book, especially when coupled with the birth of the "Star Child" at the end. Truly, the idea that "Man is a rope stretched between the Ape and the Superman", is one of universal appeal.

In religion, one of today's growing faiths is the Bahá'í Faith. Founded in Iran by their Prophet Bahá'u'lláh in 1844, they were reduced to two adherents in 1862 by massacre, but have now reached every country on Earth, with somewhere between 5 and 50 million adherents. They believe that all the religious prophets-

Zoroaster, Moses, Buddha, Jesus, Mohammed and Bahá'u'lláh, are manifestations of one God to an evolving human consciousness, and that Humanity's collective spirit is evolving towards a higher state. In this view, all antagonisms between races, nations, sexes, and religious creeds are unjustifiable errors, which diminish a person's humanity, since all people are of the same body and spirit. This Faith, perhaps alone of the world's great religions, has no difficulties in facing up to the challenges of scientific discovery. They alone also accept, explicitly, that their founder and Teacher, Bahá'u'lláh, is not the last word, and that other prophets will come later.

I make no apology for these diversions into religion since, with mythology, philosophy and science, these represents our attempts to understand ourselves, our world, and the relationship between the two. The Prophet Bahá'u'lláh, incidentally, is the first and only religious leader to have stated expressly that God has created other lifeforms of varying appearances and cultural attainments on other worlds, and my experience of Bahá'ís is that they welcome the positive use of space technology to bring humanity together and solve the material problems of Humankind, as well as welcoming the scientific insights to be gained from the New Astronomy. The idea that a space-faring world civilization could be a material substrate for the spiritual Kingdom of Bahá is not unwelcome to them, and a very powerful one for the coming Millennium. Their recent Chief Guardian, Shoghi Effendi, in "Questions and Answers", prophesied that in the coming Bahá Age Human civilization would extend into space, while Bahá'u'llah's Son Abdu'l Bahá saw Religion and Science as twin wings of the Dove of Truth.

Looking more from a biological and evolutionary point of view, the late Arthur Koestler also considered the incompleteness of Man in the "Heel of Achilles". He described how Man has evolved very rapidly in 3 million years, largely by super-imposing an enlarged cerebral cortex, with its higher intellectual functions over the primitively emotional mid brain and a reptilian, homeostatic hind brain. In saying that our hearts rule our heads, or "we know what we should do but persist in doing the opposite", we are describing a conflict between the intellectual cortex and the emotional midbrain. In his view the human predicament is that we have evolved so fast that the powerful cortex has not grown sufficient neurological connections to control the emotions of the mid brain, which are often able to colour the functions of the fore brain unduly.

The solution, as he saw it, is that Evolution will remedy this defect so as to ensure the survival of Humankind, or that we could carry out either genetic engineering or brain implant surgery to increase the neural connections from cortex to midbrain. The necessary neurosurgical or genetic engineering expertise is not yet available, but will be within a few centuries, if this is perceived to be correct. Arthur Koestler's "Greater Man" would be very intellectual, even spiritual, in full control of his emotions, but not robot-like or cold, divorced from benign feelings. Aggression and acquisitiveness would be harnessed into higher emotions such as the curiosity of knowledge and exploration, the competitiveness of sports, games, artistic creation. Nationalistic and racial feeling would be despised and checked, much as the desire for cannibalism and human sacrifice is in most societies. This person would correspond very much to Plato's Philosopher-King, or Teilhard de Chardin's Omega Man, or to the religious concept of the Good Person, blessed in the Sight of God. It may be said that such a Man would be dull without his vices and baser pleasures; these, however, are mainly trivial at best, or destructive at worst, and, on the larger scale, are likely to terminate our civilization pointlessly, whether through outright wickedness or mere carelessness. The follies of sectarianism and national cum racial hatreds are entirely base animal emotions dressed in higher clothing, and their extinction would cause us no detectable loss.

Secondly, Humankind would be developing in accordance with the higher side of its nature, which already exists-the result would not be an artificial freak.

Thirdly, the old baser "pleasures and vices" would be replaced by newer, more positive values. Lust would be replaced by love, aggression by peaceful debate and competition, bravado by a search for knowledge and adventure, and irrational hates would become redundant, as they are today at the deepest level. In losing all the-Isms which have poisoned this century, we would have lost nothing of note, and gained an eternity.

St.Paul put things well when he wrote "When I was a child, I clung to childish things: now that I am a Man, I have put them aside". Similarly, we should not miss the seven deadly sins, and our preposterous national, religious, or sexual antagonisms and hatreds if somehow the Greater Human could be evolved.

These two ideas, that of Human improvement and his relationship to space and the Cosmos, are ancient, universal, and very deeply rooted. There are grounds for believing that we are on the verge of realizing these ideas in practice most dramatically when we embark on the infinite ocean of space and adapt, biologically and psychologically, to the new demands and opportunities which will be placed upon us.

One can summarize these two chapters both by an image from nature, and from mythology.

In the Insect Kingdom, butterflies and moths are perhaps most admired for their beauty, and people have always been fascinated by the symbolism of the caterpillar pupating and emerging from the chrysalis as a butterfly. There is an obvious, optimistic, and appealing parallel between the evolution of Life on Earth and its maturation into a space-faring civilization, and the story of the butterfly. The caterpillar emerges from its egg, and devours everything in sight-plant, leaf, even its own eggshell, until, finally, it can gorge no more; it then assumes a pupal condition inside which every tissue and organ is broken down in a crisis of demolition, and is then refurbished and rebuilt over a long winter. Then Spring comes, and a beautiful, winged, butterfly hatches out which lives and feeds in quite a different manner from its ugly and greedy caterpillar , and spreads to other plants and fields to spread the seeds of life.

In parallel fashion the Egg of Earth has hatched innumerable species, and, most recently, the voracious and insatiable Homo Sapiens, which has devoured and consumed many of Earth's long-husbanded resources. We are now in a time of crisis, even like the pupa; but we stand on the verge of a brand new mode of existence, with vast new opportunities for economic and evolutionary development. Eventually, like the butterflies, we shall fly off to new pastures-the stars themselves.

Finally, to close this exploration of mythology and philosophy, there is the example of the oldest Creation myth of all. Classically, this involves the ripening of life in the Womb of the Earth Mother by the rays of the paternal Sun-God. Mother Earth ripens the seed of Life, and nourishes it to growth and maturity. The increasing evidence that the sperm-like comets deposited the seeds of life on Earth enhances and updates this ancient myth. What could be more natural, and more consistent with this archetypal feminine view of the Earth, than to see, not an Earth polluted by Man's depradations, nor a race on the verge of extinction, but rather a Mother Earth in crisis-even, perhaps in labour, after a 4 billion year pregnancy, with delivery at hand?

Chapter 3

To the Ocean's Shore.

The time has now come to consider how the twin forces of evolution and the human spirit have led humankind from the shores of Lake Turkana some 3 million years ago, to the edge of the ocean of space in the last 100 years. The joint action of human social and communicative powers and the powerful driving force of aggression led to the emergence of tool-working, weapon-wielding clans in ancient times. It is believed that people were faced with changing ecological conditions during the late Pliocene Age in East Africa. The trees and vegetation in the land of origin of early Man became scarcer owing to climatic changes, which were the precursors, perhaps, of the Ice Ages. Conditions became harsher, and people came to depend more on the use of stones and clubs to capture and kill prey, and also had to roam further afield for their roots and berries. These wanderings must have placed a premium on adventurous groups of skilled hunters who were competent at planning journeys, and attacking prey in concert.

During these millennia, hunter-gatherer groups roamed in bands covering most of the surface of the planet by the end of the Ice Ages. Soviet archaeologists found evidence in Siberia that tribes reached America from the mainland of Asia at least 30,000 years ago; Australian aborigines reached their continent from the mainland of Asia at least 100,000 years ago. This colonization of the Earth from an East African site was clearly a response to the challenge posed by our ancestors' need to pursue scarce and nimble food, whose supply depended precariously on the climate. During the last million years, the Ice sheets have alternately expanded to cover a third of our planet, and receded again during relatively mild interglacial ages. After the last one, the race of Cro-Magnon Man gradually gave way to races of people indistinguishable from our contemporaries.

The origins of the "Agricultural Revolution" are lost in antiquity, but lie between 10,000 and 20,000 years ago, at about the time of the final melting of the Ice. Evidence for the cultivation of barley has been found in Egypt dating about 17,000 years ago. During the next few thousand years, as the climate became milder, people in the Middle East, Indus and Yangtze valleys, and the Yucatan Peninsula gradually abandoned the pre-agricultural, hunting mode of existence and became tied to crops, irrigation needs, fixed dwelling places and, eventually, cities. For a long time, one of our main foodstocks was the ubiquitous mammoth, who became extinct soon after 10,000 B.C. The challenges and opportunities offered by the new way of life allowed specialization of labour, and civilization. Fundamentally, agriculture and animal domestication led to a rapid increase in human population, and to sophisticated modes of administration, and to the recording and measurement of time. Over the 10,000 years up to 1,500 A.D, the population increased 100 fold, and the needs of societies became ever more complex and interlocking. To obtain necessary or valuable materials, such as tin, spices, herbs, different crops, and animals, people had to travel further into unknown lands in search of them. When Alexander set out to conquer the Persian Empire, he took with him philosophers and naturalists to analyze and catalogue strange fauna and flora to be found in far-off places like India and Afghanistan; and, no doubt, many plants were brought back to different parts of the Hellenistic Empire.

Minoan merchants of the 2nd millennium B.C were familiar figures all over the Mediterranean, and it is thought, beyond. Legends of Jason and the Argonauts, and the voyage of Theseus, the flight of Daedalus, show the early Greek passion for exploration and trade very clearly. Exploration has often arisen from a love of adventure, curiosity and the search for some material or resource not readily available at home. Denial of such resources by others has often led to wars of conquest by trading nations. The rise of the Babylonian, Roman and British Empires was at least in part due to the need to protect trade routes, or local governments prepared to maintain them.

Discovery-Challenge and Response.

During the so-called "Dark Ages"" and Middle Ages the nations of Europe faced the challenge of interdiction of the spice and silk routes which had existed in Roman times. This blockage occurred because of the rise of a hostile power, Islam, in the Middle East and Africa, thus leading to a denial of resources in Christian Europe. The Emperor Charlemagne, who united a large part of the old western Roman Empire, was only able to celebrate the fact with silver tableware, and had no access to silk, as well as being unaware of the wider world. This is in marked contrast to his contemporaries Haroun al-Rashid of Baghdad, and the T'ang Emperor of China, who knew an entirely more imperial level of sophistication. Large sections of classical Graeco-Roman knowledge were lost until the Renaissance; this resulted in part from the sacking of hundreds of cities by waves of barbarian invaders, beginning with Attila the Hun, and ending with the Mongol hordes of Genghiz Khan in the 13 th Century. As a result of the conquests of the Great Khan, it was possible to travel safely from Vienna to the Pacific, without an armed escort, for almost 100 years; this is probably the only occasion in history that such a journey could have been made in comparative safety! However, facing the mighty obstacles of the Arab Caliphates, and then successively the Kurds of Saladin, the Mamelukes, and the Seljuk Ottoman Turks sworn to destroy the Infidel idolators of Christ, mediaeval men found the East unattainable.

Merchants and navigators of Europe, fired by the tales of Marco Polo of Venice, and many other wanderers began looking for maritime routes to the new frontier of Cathay. In similar fashion, the Vikings, had opened up trade routes between Scandinavia and America; one of their most intrepid sailors, Leif Erikson, reached New England in 1000A.D.and founded a trading post there. While pursuing the highly lucrative amber and fur trade of the Baltic, one enterprising Viking called Rurik sailed down the Volga in the 9th Century and founded a trading post called Novgorod. A formidable legacy of these Viking traders has come down to us-the modern continental power named after Rurik-Russia.

Similarly, the voyages of Portuguese and Spanish merchants, aided by the magnetic compass, led them first to Angola and Mozambique under the leadership of Vasco da Gama and Henry the Navigator, laying the foundations of great empires. Also at this time the immigrant scholars reaching Venice, Florence, and Genoa laid the foundations of a rebirth of Greek and Roman culture, and a breakdown of the bonds of mediaeval isolation.

The most spectacular result of this renewed interest in trade with Cathay and India, and the search for new trade routes, led to a remarkably wise and far-sighted investment by Ferdinand of Aragon and Isabella of Castille; in 1491, they agreed to a stake of $70 for an obscure sailor who believed that he could sail around the world to India and Cathay, approaching from the other side of Asia. This would open up a new trade route without fear of harassment by the Turks. Many members of a Royal commission set up to consider this proposal thought it far-fetched, and unlikely to produce a return on investment. In the event, Columbus did not even reach India-he reached America instead. It is well understood now, that, in pursuing the possibility of reaching India and China for commercial and trading purposes, Christopher Columbus launched a new chapter in human civilization, with results wholly unforeseen by Columbus or his paymasters.

The discovery of America, and opening up of Russia beforehand, exemplifies the unexpected and lasting results from expeditions driven by the dynamic and explosive mixture of trading enterprise and curiosity. Under these two impulses, the peoples of Europe went on during the next few centuries to explore, settle, and colonize the entire planet. It is true that much of that phase of imperialism was brutal, and opposed by indigenous populations, but it is also true that the benefits of literature, science, trade, and new cultures have made possible the modern world, with all its appalling dangers and problems, but also its outstanding and brilliant contributions to learning, and evolutionary possibilities for the future.

It is interesting to compare the response of two very different cultures to the new explorations of the 15th century.

Firstly, Portuguese and Spanish, and, later, Dutch, French and English seamen discovered new lands, developed trading and commercial relationships with them, and became wealthier and more dynamic than the sophisticated Arab and Ottoman civilizations, whose stranglehold they had originally sought to escape. These European states developed with renewed vigour, and continued to grow beyond the point where classical Graeco-Roman civilization had broken off. A scientific and technological explosion, beginning with the astronomy of Copernicus, Kepler, and Galileo led through the physical mechanics of Newton, to a new synthesis of human knowledge, which was ploughed back into marine navigational skills and chronological measurements. New intellectual frontiers accompanied the new geographical frontiers.

By contrast, however, consider the Ming admiral, born Sin Bao, (Sindbad), then titled Cheng-Po, who, in 1415, arrived in Zanzibar with a fleet of 18,000 men after an epic voyage from China, as part of a series of seven voyages. This was several decades before Vasco da Gama opened the Portuguese route around Africa. On the discovery of ivory, new spices, and new trading opportunities he returned to the Celestial Empire in great excitement, seeking to persuade the emperor to expand and develop the new possibilities. No Ming Emperor ever took up the challenge. They were the richest and most civilized Empire on Earth; what need had they of barbarian contacts at vast distance? The intrigues of domestic politics and rivalries were of far greater concern, for corruption had bitten deep into the fabric of the Ming Empire. Over the centuries, the Ming Empire languished, and fell to the northern martial Manchus the 17th century. By 1911 China stood naked before the long-nosed barbarians, weakened by decades of court intrigue and centuries of complacent decline. Finally the Celestial Empire came to a close, and China began the long march through dissolution and warfare to a new imperial dynasty. Over the 1980's and 1990's, this new China, after the turmoils of the Cultural Revolution, is reaching a new accommodation with the rest of the world, at least in the spheres of trade, science, and technology. To be sure, her economic growth since the arrival of Deng Xiao-ping and the four modernizations, has been phenomenal, to the extent that some pundits expect China to surpass the European Union in G.D.P within 10-15 years, and the United States within a generation. The Achilles heel, as with all autocracies, is the question of the succession-after the "Old Man", it is not impossible that China could face ethnic and regional civil war, or alternatively, as with many other recent examples of rapidly changing societies accompanied by increased power, a period of xenophobic, nationalistic expansion. The containment within the world economy and concert of nations of such a rapidly changing China, with all its implications for food supplies and resources, will be formidable, posing severe threats to Western based values and civilization.

The island Empire of Japan, similarly, failed to face the challenge of foreign encounters and trade during the 17th century, and remained in fossilized feudal isolation during the following 270 years. However, the Meiji restoration of a strong, modern-minded Emperor in 1870 was a response to the encounter with the American Admiral Perry in 1859; initially, this resulted in military adventurism but, when this was blocked spectacularly in 1945, this adventurism became industrial and commercial.

With the new lands, Europeans brought back not only new discoveries in geography, history and science, but also new goods and products. Potatoes, tomatos, coconuts, spices, tea, coffee, cucumbers, and chocolate are only some of the valuable additions to European diets, and hence quality of life brought about by these discoveries. There was also the powerful lure of gold and treasure, which at first enriched Spain and Portugal. However, the valour of the Conquistadores quickly gave way to the indolence of easy wealth, and by the end of the 18th Century, the gold gave out, and Spain and Portugal were increasingly left only with the fading memories of former glories; this challenge they failed to meet, and they exemplified one universal law of history and of life. A challenge unmet is a signpost to decline.

Many cultures and civilizations have met the challenges of local overcrowding, resource depletion or denial, oppression, and incessant strife by migration and settlement of new regions. I have described the experience of Columbus and the new World because it offers interesting parallels with the story of modern civilization. It also furnishes a useful precursor for the central subject of this book-the coming human migration into space.

The same story has been repeated in many lands, and at many different periods of history. Another example is the Greek colonization of the lands and islands of the Mediterranean, Aegean, and Black Sea during the first millennium B.C.; the cities of Colchis, Byzantium, Syracuse, Rome and Antioch are a few examples of a massive mercantile expansion; this was brought about by Greek maritime commercial enterprise which sought new opportunities for those excluded by narrow oligarchies at home.

Byzantium was a colony of Corinth founded in 800 B.C., which soon greatly outstripped its Mother city-we know her now as Istanbul, and for 1200 years she ruled the Eastern Christian world as Constantinople-Istanbul is simply a Turkish corruption of the Greek phrase-"eis thn polin"-which means simply "to the City". For well over a thousand years "the City" meant Constantinople. The Greek colonies were the product of a classical mixture of business enterprise, adventurism, and local social and political factors. Each colonist no doubt acted from personal and local motives, but the resultant of all these individual choices was an enduring Greek tradition of trade and marine commerce which survives to this day.

A similar mix of motives, with the added one of a conscious desire to remake the world as one vast Hellenistic civilization, led Alexander the Great of Macedon to subdue the Persian Empire in 333-323 B.C. There was a scientific and exploratory dimension to this epic; he took with him a company of scientific advisers and philosophers. Many cities and colonies were founded by Alexander, notably Alexandria of Egypt, which, up to the Christian burning of A.D.300, contained half a million priceless texts of ancient knowledge and literature. Even today, over 3 million Egyptians have demonstrable Greek ancestry and practice the Coptic form of Christianity, and, in the frontier regions of North-West India, Iskander arrived only the other day! If Alexander had stayed at home, and Ferdinand and Isabella of Spain had not backed Columbus in 1491-2, their names would have remained sunk in a well-deserved obscurity.

Rapa-Nui-our Earthbound future?

In the Pacific, the settlement of Polynesia and Micronesia bear witness to similar motivations. The central actors remain unknown, but in the first millennium A.D. heroic feats of navigation across uncharted ocean deeps led to the settlement of islands as much as 3,200 kilometres from land-most notably Rapa-Nui (Easter Island) in about 400 A.D. We do not know who these pioneers were by name, in most cases, or what drove them; but it seems likely that similar universal motives drove them to seek new homes and fortunes across the seas. Recent research on Easter Island has unearthed oral traditions of their foundation and ultimate fate. A minor chieftain named Hotu Matua was driven by population pressures and clan rivalry to set off from the Marquesas Islands to sail ever further East across the Pacific, and finally reached Easter Island, the last island before Chile.

The riddle of the erection of vast stone statues, clearly by a large and sophisticated population over a period of centuries on a treeless island, with no visible means of levering such colossi into position, has led to some pretty wild speculation about the origins of the Easter Island civilization. It now transpires from recent research by palaeo-botanists and archaeologists that the problem is entirely understandable, and from a modern point of view, chillingly apposite. Early in the first millennium,

Easter Island was well wooded with Chilean palm trees, and fertile soil. The first settlers congratulated themselves on finding good land to settle, even though it was 4,000 kilometres from home. Over the centuries, their numbers grew to somewhere between 40-50,000 sustained by fishing and intensive agriculture, and, in homage to their ancestor-gods, they built stone statues, to guard the edges of their world. For many centuries, they believed, their island was the habitable universe. From 1100 to 1700 A.D., they built ever larger statues, felling trees to provide the necessary rollers and levers. Finally, in the 18th century, the inevitable happened-they began to run out of trees, and so could not only not continue to build statues, but could not go deep sea fishing either. By the time they realized their plight, they had lost the arts of shipbuilding and navigation and found themselves trapped on their remote island. Between the brief visits of Roggeveen in 1722, and Captain James Cook in 1774, they changed their gods, casting down the clan ancestors, and worshipping the Frigate bird, in what was surely an unconscious realization that the Frigate bird had the only viable solution to the predicament of the islanders-flight! They took up arms, and fell into a generation of starvation and internecine civil war which carried off at least 60 % of the population, leaving a few emaciated, "shell-shocked" survivors on a barren island to greet a mystified Captain Cook. There has been a slow recovery of economy on the Island with the advent of alien western tourism, and, in what must surely be a supremely symbolic gesture, NASA has built an emergency landing strip for the Space Shuttle. Easter Island is an enduring object lesson in the effects of over-population and resource depletion on a finite area. For Easter Islanders alone, as they thought, in all the Cosmos the end must have been unbearable.

Earth differs, of course, from Easter Island, in two respects-it is bigger-though still very finite-and, unlike Easter Island, the resources of the solar system are more easily obtainable than the outside world was for the unfortunate islanders. We do not have the excuse that outside resources are beyond our ken or our reach. We also have, in good time, arrived at the necessary means. The story of Easter Island gives us little ground for believing that we are likely, or even constitutionally able, to cheat fate by forbearing from "felling the last tree", whether literally or figuratively. There are those who seek a future without the resources of space, or the need to build an extraterrestrial civilization. Their chosen future is on Easter Island, and it does not work!

The Viking voyages of AD700-1000 attest similarly to drives of curiosity, commerce, and the search for new resources in new lands as universal drivers. These characteristics are essential to humankind, and, when denied or repressed as in Ming Dynasty China, pre-Meiji Japan, or Mediaeval Europe, lead to stultification of the spirit.

In the present era, how are such drives to be satisfied without warfare and strife, since the world is now a completely inhabited gold fish bowl. Adolf Hitler sought new lands and opportunities for the German people with all too dreadful results. The Nazi era proved that there is now no room for the expansionary, conquering spirit of Humanity on this Earth which is compatible with the survival of civilization, so that we have reached an impasse-a classical situation of Toynbean challenge.

I has been shown how, in different times and places, people have taken up the classical Challenge which the late historian Arnold Toynbee ascribes to History, and how some have responded and developed, while others have failed to meet the challenge correctly, to be rewarded with decline or destruction. This law would appear to be ineluctable-nowhere in Nature are there grounds for believing that static equilibrium can persist forever. There is only expansion and development, or contraction and decline. An apparently static condition is only an imperceptibly slow stage on the path to decline, or a pause before further growth.

Let us now look at the stages of the evolution of America, as these are well charted, and show the general dynamics of humanity's encounter with new land or resources. It is at least reasonable to assume that our future in space will follow similar laws, as long as we are human, and the history of space development so far can be clearly shown to demonstrate these dynamics in action. We shall then consider the present world situation, and see how clearly it fits the pattern of a gigantic Toynbean "Challenge" meriting an equally radical and massive "Response".

The settlement of the Americas over the past 400 years exhibits four classic stages in the development of any new resource, geographical or economic. These can be seen with lesser clarity in other similar situations such as the Greek colonies, especially Byzantium, and the development of Southern Africa and Australia.

The first phase we call that of Discovery and Exploration. In the beginning, the lands are either unknown altogether, or their existence only suspected-they are reached only by extreme daring, and against the well-nigh universal conviction that their would-be explorers are quixotic lunatics at best, or confidence tricksters at worst.

Columbus, for instance, took a long time and much intrigue to persuade the court of Spain to provide him with $50-then a considerable sum-to fit out an expedition to search for India and China, and re-open lost trade routes. It is hard to believe that there should have been such difficulty in obtaining $70 as a reasonable investment, but it is to Ferdinand and Isabella'a undying credit that the money was forthcoming and that they over-ruled their more prosaic advisers. In the event, his expedition and others in the decades that followed discovered two new continents, and proceeded to explore them. The early settlements in Central America and the Caribbean were the first sites of this exploration, and within 30 years Pizarro had discovered the vast Pacific Ocean on the other side of the new continents.

In 1497, John Cabot, likewise, approached his sovereign, the canny Henry VII of England, for support; Henry "invited" his bourgeois and merchant classes to find the money. Faced with Cardinal Morton's infamous "taxation fork", and the Court of the Star Chamber, they complied, and England attained the discoveries in North America for $28. In today's terms, a similar effort ($28 millions) by the UK government could launch the spaceplane Skylon into service, with similarly incalculable long term benefits!

During this first century, Magellan, Drake, and many others discovered the true extent of these new lands, and went on to circumnavigate the world. The true size of our planet began to be appreciated, and a global consciousness became possible. New civilizations-Mayans, Incas, and Aztecs were discovered and, alas, subdued by armed force and new diseases. During these voyages, new medical problems were identified, for instance, scurvy. This resulted from the absence of fresh fruit containing Vitamin C on these journeys, and cost the early explorers dear. Magellan lost at least one third of his crew from scurvy, and this remained a problem until well into the 18th century. It was learned that citrus fruits, notably oranges, lemons, and limes, if taken as a regular addition to the diet, would prevent this disease, thus greatly extending the range and safety of long-term sea voyages. Advantage was taken of this knowledge by Captain James Cook in his epic 3 year voyage of exploration and discovery, which brought into consciousness the fabled Southern continents of Australia and Antarctica, as well as the South Sea Islands, not least Easter Island.

Arthur C.Clarke has compared these oceanic voyages in time scale and complexity with the interplanetary voyages to come-these will also be of months or years duration, involve small crews in close quarters, and will only be safely made as a result of greater understanding of human physiology and medicine in extreme conditions. Sailing ships on these voyages had to be largely self-sufficient, with only rare stoppages at harbours to take on new provisions and crew. The ships' artisans had to be very skilled at executing "running repairs" or "making do and mending". We have already seen in Apollo 13, Skylab, Salyut 7,and, more recently, Mir heroic examples of these same qualities.

The second stage, exploitation and commercialization, followed a few decades later. It was soon found that the Americas were not just geographically new, leading to new ideas of this planet and of cosmology, but also packed with new materials and resources. The Americas produced the new foods-potatoes, tomatos, paw-paws, avocados, tobacco, coffee, and cocoa ; many of these were imported into Europe and led to startling improvements in diet, with consequent gains in health, nutrition, and population. The bark of the cinchona tree, containing quinine became a new addition to the pharmacopeia. New trade patterns were opened up, including the wicked triad of slaves from Africa to Europe and America, to work sugar plantations in the Caribbean, which brought revenues back to the exploring nations. In the case of South America there were also the vast empires of the Aztecs and the Incas, with their golden riches, which lured in such adventurers as Cortez and Pizarro. There were also mines of emeralds, diamonds, tin and iron over the next century which added to the wealth of Europe.

During the latter half of the 16th, and early 17th centuries, small settlements of colonists began to spring up all over the Americas, from Hudson Bay in the North, to Peru and the Southern regions of Chile and Brazil in the South. This was the opening of the third stage, that of colonization, which led to the settlement of the new lands by thousands of Europeans-Spanish and Portuguese at first, followed by Dutch, French and English.

Their motives were mainly commercial-to trade in the new commodities-furs, crops, whales, gems, ores, fruit and herbal remedies. It was necessary to establish trading posts to handle the collection of merchandise and ship it back to the mother-countries. Examples are the Hudson Bay Company, which began life trading in furs, and, on the other side of the world, the East India Company, which traded in tea and spices. There matters might have rested had there not been other factors at work.

During the period 1500-1700 it became apparent that there was a new, more self-reliant, enterprising life-style to be made by individuals strong and resourceful enough to brave the voyage. For those willing to overcome the hazards of opening up a new frontier in plantations, or trading in a far-off, often hostile continent, the freedom and riches of a New world were there for the taking.

Many restless young Europeans blocked by relative over-crowding, and sickened by incessant religious warfare and persecution, sought material opportunities and conscientious freedom in America. The voyage of the

Mayflower in 1620 exemplifies this, and is regarded by many Americans as the true genesis of modern America. These were the first of millions who were to settle in America as colonists, not necessarily in search of gold or wealth, but of freedom from persecution, bigotry, and limits imposed by traditional European societies, with their rigid class distinctions. The French colonists of Louisiana and the southern States sought new opportunities to own land; at home they were cramped by the rules of inheritance, which meant that successive generations of nobles had to share out smaller estates among more younger sons. The emerging middle classes had very restricted opportunities under the "Ancien Regime".

Protestant sects sought refuge from the Spaniards and the French Catholic monarchy, while Catholics sought refuge from Elizabethan, Orange, and German persecutions. Jews sought refuge from one and all. In America, they were able to establish communities based on there own radical ideas, and to live in tolerance with other, more differently constituted communities. There has, indeed, been a succession of waves of refugees to America from different persecutions and hardships-French Huguenots, English Catholics, German Protestants and Catholics, Dutch Huguenots, Russian and east European Jews from the Czars and the Nazis, Irish from the potato famine, Balts from Soviet occupation, and other refugees from the age-long German-Russian struggles over Europe.

These have led to the formation of a new, vigorous, and radical country whose characteristics are not simply a copy of its European constituents. By the late 18th Century, it was apparent that the colony was no longer an appendage, but capable of self-sufficiency and independence. In 1776 this became an established fact with the Declaration of Independence, and the United States of America was born.

The fourth stage, that of civilization and expansion, began, leading to the settlement of the entire continent. During the years 1800-1900, America became not merely independent with its own history and values but also more powerful and vigorous than the collection of European states from which it sprang-so much so that it was able to help rescue

Europe from dire straits on two occasions, and make its own unique contribution to human destiny. Like the colonies of Byzantium before her, the U.S.A came in time to overshadow the states of her origin.

Similar stories have unfolded in different lands during the past half millennium-Australia, South Africa and New Zealand have similarly grown away from humble beginnings to became individual, independent states with their own contributions to the world mosaic. India and the rest of Africa had a less prominent Stage Three than the Americas, while South America's stage four has not been so successful as yet. The sequence of Discovery, Exploitation, Settlement, and Expansion is a useful model, and can readily be adapted, in the chapters that follow, in the conquest of space.

The wheel has turned full circle; five centuries ago, the discovery of the Americas brought new vigour, wealth, land and opportunities to a constricted, over-crowded, and war-torn Europe. It is from America, now, in concert with Europe and other groups of nations that the world as a whole must expect a similar response to the crisis facing human civilization in the coming half century. The ending of the Cold War, and the final end of the space race in 1995, with the merging of the old Soviet and the American space programmes in the construction of the International Space Station , give us the hope that, in keeping with the magnitude of the Challenge, and the necessary extraterrestrial response, the breakout from Earth in the coming Millennium will be truly global.

Arnold Toynbee, the historian, saw the historical process as a rise and fall of civilizations; he saw civilizations as growing in vigour as they respond correctly to challenges thrown at them by the world. For instance, religious groups in Renaissance Europe responded to the challenge of persecution by emigrating to America and establishing their own communities and so prospered. The Romans responded correctly to the mercantile marine of Carthage by learning the arts of naval warfare ; they later failed to respond to the challenge of the Goths and Visigoths because they dissipated their military skills in fratricidal strife instead of uniting to meet the new threat.Ming Dynasty China responded to the Challenge of European trade and technology by trying to exclude the alien ideas and maintain an aloof superiority, but in the end succumbed, whereas Europe, faced with the Islamic obstacle to Eastern trade, first circumvented it, and then overcame it.

Today's Challenges and Responses

Humankind faces a crisis of an altogether different and greater degree than any to date, although, on inspection, it is not really different in kind from that faced on smaller scales by other societies. A third of humanity has a relatively affluent standard of living, with good expectations of health, education, food, transport, consumer goods and so on. All of these ultimately are the fruit of the Industrial Revolution, which took place in Europe from 1750-1850, and the rise of western civilization arising from the Age of Discovery. There are, however, signs that, with the increase in economic instability and recessions that this happy situation may not continue much longer, and that Western civilization, in addition to the global challenge of the next century, is also in danger of senescence and decline. This wave is spreading over the whole world, but has left about 1-1.5 billions still in pre-industrial conditions of poverty, isolation, and hunger. It is also apparent that the standard of living of the wealthier nation depends, in the final analysis, on the availability and cost of energy, and that most of the sources of energy now used are non-renewable-at least on historical time-scales!

We face, logically, a few hard choices. We can limit growth so as to avoid further despoliation of the Earth's natural resources. In this case, we can decide to share what there is between the wealthier and poorer nations. If so, the wealthier would have to reduce their present standard of living to one third or one quarter of their present levels. This is inconceivable in any society other than a rigid despotism, and would have to be imposed by terror. In any case, because of the growth in population, it would only elevate conditions in the poorer countries by 20-30%. A very large number of people would have to willingly accept a great reduction in living standards so that an even larger number of very poor people could become marginally better off. Furthermore, it is an absolute certainty that India and China, who outnumber the West considerably, will not accept curbs on their economic expansion in the next, critical 5 decades-not even to please the Western environmentalists! It would scarcely be necessary even to mention this possible solution, were it not for the fact that during the coming decades, various bands of ultra-green zealots will cunningly conceal a desire for despotic power under just such a camouflage.

If we decide to maintain the present disparity between rich and poor nations, this is likely to lead to wars of redistribution during the early years of the next century. The Gulf War, fuelled ultimately by a fall in oil prices engineered by Kuwait, led to a desperate and ham-fisted attempt by Saddam Hussein to evade the box he found himself in. The relative ease with which the coalition forces freed Kuwait is testament to the selective power of the new technology spawned by the West's investment in space and electronic technology. The Gulf War is a foretaste of the resource wars yet to come. These may well become more threatening when dressed up in the tinder of Religion and Nationalistic ideology. Those interested in causing disruption, whether under the disguise of Nationalism, neo-Marxism, militant Islamic Fundamentalism, or just plain love of violence, are likely to find their task facilitated in the conditions likely to be prevalent in coming decades. Small-scale nuclear warfare would become a highly probable result. We have seen Saddam Hussein come within 3-6 months of an atomic bomb, while Iran is said to be 4-5 years away now. It is now believed by some that even a relatively small nuclear exchange, of 100 megatons or so, could precipitate enough dust and smoke into the upper atmosphere to block out the sunlight from the Earth for up to 1 year. This would cause Arctic temperatures over the entire planet. As of November 1983 this effect has become widely known as the "Nuclear Winter". Human civilization would sink into eternal oblivion, and Creative Evolution would have to resume the long march to Intelligent Life with another candidate. Since 1983, doubts have been cast on the details of this scenario, the megatonnage required to cause the nuclear winter effect, and whether the cooling would be so severe. Computer models are affected by the effects of oceans and mountain ranges, which could mitigate the effects. However, there are three points that need to be taken into account;

1/ The likeliest theatre of a nuclear war today, the Middle East, has an unusual potential for filling the upper atmosphere with thick black smoke- the oil wells. The local nuclear power, Israel, has 300 nuclear missiles and bombs, and is still ruled by people with memories of the Holocaust. If faced with an Islamic Fundamentalist threat to their existence, possibly backed by a nationalistic, anti-Semite Russian government, there should be no illusions about Israel's preparedness to unleash nuclear retribution.

2/ The evidence of the great extinctions due to asteroid impacts in the past, and the climatic cooling, with accompanying hardships, which have occurred in historical times, after volcanic eruptions- notably Mount Tambora in 1815, and Mount Hekla in 1054-show the reality of global cooling.

3/ Most tellingly, we only get one chance to find out if a nuclear winter is real-if the sceptics have it wrong, it will be small consolation to those who believe in nuclear winter.

The creation of a global consciousness, together with a widespread understanding that part of the challenge facing humanity is extraterrestrial, in the form of large cosmic mountains of rock, would give us the right frame of mind with which to face the next Millennium.

We are playing for high stakes indeed!

The only practical and sane, humane, alternative to a reversion to the Middle Ages, or increasing North-South conflict leading to the extinction of civilized life would therefore seem to be a global attempt to raise the living standards of the rest of humanity to something approaching present day conditions in Europe and America.

This can and must involve a degree of industrialization or at least, energy production and consumption of water and other resources, wholly beyond today's imagination. This will imply desalination, or the transport of ice from the Pole caps. Many countries in Africa and Asia will experience drastic shortages of fresh water over coming years unless this is done. The correlation in recent studies of drought periods with the vagaries of the sun-spot cycle only attests to the complex nature of the challenges facing us, and illustrate that part of the challenge is extraterrestrial, and that our success cannot be complete if we confine ourselves to terrestrial responses.

Fuel is required to supply heat or electricity for all industrial and intensive farming purposes. At present, most of this comes from fossil fuels-wood, coal, oil, and gas, whose reserves are finite. For a world raised to European styles of living these would be exhausted well within a generation. In October 1983 America's Environmental Protection Agency and their National Research Council both warned that, if fossil fuels were used at their present rates, the atmospheric Greenhouse Effect would become visible in climatic effects by the 1990's. Now that the nineties are here, the question of whether the effect is already under way is under intensive study, complicated by the variability of solar output revealed by space research. In the greenhouse effect, a rise in atmospheric carbon dioxide resulting from fuel combustion coupled with tree felling leads to a trapping of solar infra-red radiation within the atmosphere which cannot then be re-radiated onto space after sunset. This same principle operates on a small scale in the garden greenhouse, where it is useful. It should be pointed out that Earth's greenhouse effect is not new, and that, without it, the planet would be a pretty chilly place-that problem is a matter of degree, not of kind. Applied to the planet, within 20-30 years, said the E.P.A., wind and rain patterns will shift, leaving the Southern States and Europe desiccated and desert-like. Crops would start to fail, and by about 2040, the atmospheric temperature *could* have risen by as much as 4 C, causing the ice caps of the poles to start melting. Recent satellite data shows a greater degree of "calving" of icebergs at the edge of Antarctica than at any time in recorded history. Severe flooding would affect many of the major cities of Mankind. After a certain stage, carbon dioxide will be released from solution in warmer water, and from chemical bonding with rocks and salts. This would lead to a "runaway "greenhouse effect, caused by a positive feedback loop. No-one knows at what point this latter situation would arise-except that it did indeed happen on the planet Venus, still less is it possible to make specific climatic predictions for particular areas of the Earth. One can be sure that a more rapid and violent change in climate than any in the last few millennia is bound to have dramatic and unpleasant consequence, since any change if suffered over a short period is likely to be adverse, if only because eco-systems, like humans, need time to adapt to change. All the above arguments do not take into account the continuing industrialization of most of the "Third World", let alone the fact that human population will reach 7 billions by 2010, and 9-10 billions by 2050. This will greatly exacerbate all the dilemmas raised earlier. The earlier EPA projections have been revised downwards, with the average temperature increase reduced to 2-3 degrees over the coming century, but international treaties to combat the problem, although desired, are fraught with contention. The question of energy supplies is not greatly helped by the suggestion that nuclear power, whether fission, fast breeder, or fusion, will solve the problem. In any event, since the disaster at Chernobyl in 1986, there is not the remotest chance that fission or fast breeder reactors would be built in the numbers and timescale required, even if it were deemed desirable. Fusion remains a laboratory exercise, although it advances by the decade towards the point at which useful energy could be produced. It will in any event be fiendishly expensive, unless it can be facilitated by the use of the Helium 3 reaction-which is apparently easier to achieve than hydrogen/hydrogen or deuterium/hydrogen fusion. Helium 3 is easily obtainable-on the Moon. The E.P.A's-and Kyoto's-favoured solution is simply to cut back on fossil fuel consumption, but this leads us back to where I started. In January 1998, however, it was rumoured that the U.S. Government is to make a grant of $150 millions to NASA to look again at the clean renewable option of solar power from space-of which more in later chapters.

Whither Civilization?

The Kyoto conference of 1997 reiterates the danger of global warming in the coming century, and calls for planned and agreed reductions of fossil fuel carbon emissions to avert a 3 degree rise in temperature, while the progressive destruction of temperate and tropical forests alike removes one half of the Earth's coping mechanisms. The need for careful monitoring of compound atmospheric, solar/terrestrial, volcanic and oceanic effects on our environment, using increasing surveillance from space, is now self-evident except to a handful who would simply go back to a never-never period of untechnological bliss in the Middle Ages. The need to shed 4-5 billions of people to attain this romantic goal has so far eluded them.

All the problems were exacerbated by a new element-namely, the downturn in several Western economies during the late Eighties, with the growth in apparently permanent un-employment. In 1997, several so-called Asian Tiger economies met downturns of their own. This made ideas of cutting back on growth and consumption even more unsaleable, and socially hazardous to many nations. Recent economic upturns have improved this situation, largely due to the emergence of new technologies, as described by the cyclical economic theories of Kondratieff. He proposed in addition to the well known 10 years cycles of growth and recession a larger longterm 50-70 years cycle of growth due to the entry of groups of basic new technologies (steam and coal , oil and automobliles , for instance) which allow complete restrutcting of societies. Some see the space industries as part of a new complex of technologies which will usher in a new period of growth and change in the New Millennium.

It is difficult to conceive of a continuing spiral of Kondratieff style growth cycles by technological revolution being confined to one planet without endangering our natural environment.

The last great recession in Western Europe, in the 1930's, was ended by the frantic rearmament drives occasioned by the advent of Hitler-whose rise was not unconnected with economic hardships. We have an interesting conundrum in the post Cold War era, in which the armaments industries, without a clear "enemy", are in recession, and thereby threatening to spread the waves of recession in the rest of the economy. Armament industries can only survive nowadays by supplying weapons to their countries' potential enemies. For Russia's part, unemployed armaments workers and scientists are in real danger of being lured to unreliable regimes in the Middle East, while the Russian government finds itself driven to sell nuclear reactors to an oil-rich neighbour, in the full knowledge that nuclear weapons will be built there and, as likely as not, used against Russia herself.

The world thus badly needs a new "economic multiplier" , of a high-tech nature, that will free us all from the crazy dance of death with armaments, to provide for economic prosperity in the next Millennium, and which does not wreck the planet's ecosystems. It would appear, then, that within the next two to three generations-70 years or so at the most-the ecological crisis of population-energy-resources-food and environment is set to become a Toynbean challenge of massive proportions. Unlike previous ones, it is global in scope, and threatens the very survival of many higher forms of life on this planet. It is thus an evolutionary as well as a historical crisis. It also appears that terrestrial solutions lead either to rapid ecological disaster, or to a slow decline into an overpopulated squalor accompanied by diminishing dignity and freedom for all. Even if populations were somehow to be curbed without tyranny or worse, the result would be a global senescence, leading ultimately to spiritual and cultural decline. Humanity cannot fulfil its higher potential with a Zimmer frame. Civilized Life bound to the Earth will either not survive the next few generations, or will be unworthy of the name. I prefer to call the Earthbound solution to our problems the "Easter Island" option.

A response to this Toynbean Challenge will need to be massive, radical, and already visible if it is to be timely and appropriate. Toynbee himself proposed the emergence of a Universal World Order based on Christianity as the sole response worthy of the Challenge posed by the modern world.

However, on its own, religion would not answer the very real economic and material aspects of the situation-the religion would need, as the Bahá'ís would be the first to accept, a material substrate. Toynbee did not live long enough to see the Response that humanity is already preparing. Few are aware of it as yet, but a Response is in embryo, and growing. It is entirely an natural and understandable one in the light of the evolution, mythology, psychology and history of our species as outlined in this book. Indeed, it can be seen in retrospect that both the Challenge and the Response have been long in the making. It will be a classic one of the discovery of a new Frontier, exploration and development of its resources and properties, settlement, evolutionary adaptation, and, eventually, limitless growth and development of Humankind.

We have arrived at the challenging conclusion that there is no Earthbound future for humanity which is compatible with freedom or dignity. We have grown too big for confinement on one planet, however beautiful. The correct, truly Toynbean, Response is now obvious for all to see, and its origins, development, and realization will be the subject of the remainder of this work-such an Earthbound future as is now discernible demands an extra-terrestrial Response.

We have reached the shores of the Ocean; it is now time to set sail upon its waters.

Chapter 4

Exploration and Discovery-The Voyage Begins

The birth of the Space Age was a long time in coming. Serious or semi-serious proposals to travel in space began to appear in the mid nineteenth century. For it was only at this time that people began to dispose of sufficiently powerful propulsive forces, strong enough materials, and adequate skills in navigation even to consider such a venture.

Before the advent of high altitude aviation and sounding rockets, the nature of the atmospheric upper regions was little understood, and , before Sir Isaac Newton promulgated the laws of gravitation and inertia, the nature, and distance of the planets were hardly guessed at. To conceive of a flight to the Moon in all seriousness, it was required to show that that body was in fact a body, with physical features, which could in theory be attained by physical means.

At first, indeed, because the nature of the upper atmosphere-its temperature and density-was unknown, it was thought that people could simply fly to the Moon, like a bird, through the air. Cyrano de Bergerac in 1650 was imagined as being drawn there by the power of evaporated dew, while other would-be travellers were to be pulled there by swans. In 1600, the Chinese warlord Wu fitted rockets to his chariot, and suffered a "launchpad explosion"-not the last of its kind.

Ironically the true instrument of spaceflight-the rocket-had been around for several centuries, being known to the Sung Dynasty of China in 1200 A.D. It was used for ceremonial occasions, however, and its unique properties of propulsion were not appreciated. The fuel used follows the same principles as all chemical rockets up to and including the Space Shuttle and Venturestar; in all rockets, a chemical reaction takes place between a fuel (and an oxidizer to produce hot, expanding clouds of gas which, if ejected through a nozzle, will push the rocket forward in the opposite direction. Gunpowder is not a very potent fuel. The gases are neither hot nor active enough to produce enough thrust for our purposes, as the powder does not burn evenly or cleanly.

The conceptual foundations of spaceflight were laid in the years between Copernicus and Newton-1480-1680; during this era Space became a place, and the planets became worlds. With Newton's celestial mechanics their movements became predictable, and their distances and sizes became known. It is Newton's Third Law, however, which underlies all of astronautics. This law states that, for every action there is an equal and opposite reaction.

The Rocket Equation is expressed simply by the formula
$Vf = Ve*2.3 \, Log10*(M1/M0)$,
where Vf =Final speed of the vehicle, in Kms/second
Ve=speed of the exhaust gases, in Kms/second
$Log10$ denotes base 10 logarithms,
$M1/M0$ is the mass ratio, or mass of vehicle fuelled, divided by Mass after fuel is burnt; this, in effect, is the payload plus the structural mass.

It is this relationship which governs the performance of all present forms of rocket propulsion. All combinations of fuels, electric ion engines, mass driver systems, fusion , fission, or anti-matter engines, are all methods of increasing either the mass ejected per second, or its speed of ejection. Another quantity, the Specific Impulse, is a measure of the time for which a given amount of thrust can be derived per kilogram of any given fuel-typically, in a Space Shuttle Main Engine, this is about 460 seconds-the best that can be done with conventional liquid fuelled engines. Nuclear engines, in which a fuel is heated by an atomic reactor rather than a chemical reaction, could top 900-1000 second specific impulse, during tests in the early 1970's. Exhaust Velocity, *Ve*, is determined by the reaction temperature, and by the lightness of the fuel, or, in maths, $Ve=T/M$, where T=Temperature, and M=molecular weight-hydrogen being the lightest.

Much current effort is going into reducing the structural component of the M0 term in the Mass Ratio-new lightweight materials, such as a new Russian lithium/aluminium alloy, epoxy/graphite, and kevlar analogues are being developed in the Delta Clipper and X-33/X-34 programmes, and, with lighter electronics, promise to bring the Holy Grail of Single Stage to Orbit, reusable vehicles to reality in the next few years-more on this later. In one sense, the above mathematical considerations, and underlying physics, apply to any imaginable feat of space propulsion, and any new space vehicle encountered by the reader (in the foreseeable future) will perform exactly according to these rules. No new laws of physics will be required to undertake any of the programmes to be described herein, although their application may take ingenious new forms!

Newton had also established the laws whereby any object accelerated to a sufficient speed would eventually leave the Earth. This became known as the escape velocity, and, because the Earth is a rather massive object, turns out to be very high-40,000 kilometres per hour (11 Kms/second). The famous example of the apple and the tree was not merely the discovery that apples fell down. The significance lay in the fact that, if blown by winds of different speeds, they would land at different distances from the tree. It followed that, if a sufficiently strong wind were to cause the apple to travel fast enough, it would travel so far from the tree as not to land at all; it would, in fact, orbit the Earth. This speed, which the Russians call the First Cosmic Velocity, is about 8 Kms/Second, or 29,800 kph. His mechanical laws enabled people to work out that, at this speed, the apple would orbit the Earth at an altitude of just over 160 kilometres, and become a satellite of the Earth. He carried this reasoning a stage further, and supposed that the Moon might be a large-scale example of such an apple, falling for all eternity towards the Earth, but moving at such a speed and distance as only to succeed in circling it. Escape velocity, 40,000 kph, is the Russians' Second Cosmic Velocity.

Interestingly, the escape velocity from the Sun's influence is 60,000 kph, so that, at that speed, a vehicle can leave the solar system altogether, and travel through interstellar space-the Pioneers 10 and 11, and Voyager 1 and 2 probes are our first examples. It was this fact that led the late Robert Heinlein to remark that 160 kilometres is half way to Infinity, since to accomplish any voyage in space, it is the *Change in Velocity*, or Delta V, which determines what can be done.

Visions of Space Travel.

It was not long before scientists and novelists began to see the implications. The astronomer Edwin Hale in 1869 wrote a novel called "Brick Moon" in which an advanced race had built an artificial satellite, and astronomy could be pursued with a new vigour. Jonathan Swift's satire "Gulliver's Travels" also includes a description of the flying Island of Laputa, where new stars,10,000 of them, could be seen with Galileo's new telescopes.

The balloon flights of the 18th and 19th Centuries, beginning with the brothers Montgolfier in 1783, opened man's conquest of the air, but, paradoxically, greatly diminished the dream of space travel. For it was soon found that the atmosphere became thinner and colder as balloonists rose higher. Flight of more than a few kilometres above the Earth seemed if anything more remote than ever. The attempts to achieve heavier than air flight during the 19th Century, using imitations of birds' wings, and early gliders, produced some heroic feats. Sir George Cayley and Otto Lilienthal achieved great successes-the latter before his death in a 15 metres fall. They did little to help spaceflight, so the dream of interplanetary travel became more remote. It was realized that the obstacles were simply too great for the times.

The dream, however, did not die. Jules Verne, struck by the success of the Yankee industrial machine, with its armoured steam-powered gun-boats and heavy breech-loading cannon , began to wonder whether here was a new industrial power, possessing great inventive genius, which might shoot for the Moon. In his "Voyage Round the Moon", in 1870, he gave the first reasonably scientific account of a flight to the Moon. 3 men were launched in a capsule from Florida on a journey to fly past the Moon, swing round it, and return to a "splashdown" in the Atlantic Ocean. The effect of Earth's and Moon's gravitational interactions, the effects of weightlessness, and the close passage of the Moon are described very vividly, as was the necessity to carry oxygen and carbon dioxide purifiers on board. Also well treated were the concepts of orbital and escape velocities, and the inability to change direction without an on-board propulsion system. The voyage was an excellent foretaste of the real thing in December 1968. This was the immortal voyage of Apollo 8, when Frank Borman marked Christmas Eve with a memorable reading from Genesis against the backdrop of a lunar landscape. That voyage showed us Earth in its true perspective for the first time, as a fragile, small, blue planet against the background of limitless space, with the Moon in the foreground. Since then, our consciousness has passed other milestones, with the historic picture of the double planet Earth-Moon system shot by the probe Galileo 11 million kilometres on its trip to Jupiter, and the view of a small blue single-pixel dot on an inky background, seen from outside the solar system by Voyager, on its way to the stars, past the orbit of Neptune.

For all that Jules Verne foresaw, one can immediately contrast the event from the story ; Verne's crew were hurled to the Moon, not by the slow graceful, acceleration of a mighty rocket, but by the terrible accelerating shock of a vast military cannon. The vehicle was not a true spaceship, but an artillery shell. This was the most powerful propellent available to Jules Verne's imagination, and totally unsuitable. The crew would have been crushed to death instantaneously by the accelerating forces of the explosion, even had the vehicle survived intact. The absence of radio communications and of computerized navigation were the other two major features in which reality differed from fiction. Verne's vehicle was the work of an amateur organization, the Baltimore Gun Club, not the vast state machine of N.A.S.A. The first true theoretician of spaceflight was the Russian Konstantin E. Tsiolkovsky, a mild mannered school master, born in Kaluga 110 kilometres South East of Moscow, who wrote of travel to the planets in an age where most of his fellow countrymen travelled by horse and cart. He is, as far as is known, the first author to consider seriously the true key to space travel-the rocket. He saw that this reaction engine alone could achieve sufficient velocities to reach the upper atmosphere and beyond. He saw plainly that the gunpowder rockets of the pyrotechnicians of his day, and the larger military rockets used in the Napoleonic Wars by William Congreve, could, in principle, fly in space if sufficiently energetic fuels could be developed. Indeed, as they work simply by Newton's Third Law, they would perform more efficiently where there was no air resistance to slow them down. He also realized that, if a rocket tube were fitted with fuel so that most of the mass at launch was in fact fuel, the rocket would slowly increase in speed, as the consumption of fuel reacts against an exponentially decreasing mass-n.b. the "rocket equation" derived previously. This is in sharp contrast to Jules Verne's artillery shell, which is kicked out by a tremendous explosion, only to lose speed as it meets air resistance on its way up through the atmosphere.

Tsiolkovsky also realized that a rocket, because it carries both a fuel, and its own source of oxygen will work just as well if there is no air for burning the fuel. Indeed, in the vacuum of space, the rocket is in its true element. He described his rockets as being powered by as yet unspecified liquid fuels, which would contain two elements-a hydrogen-rich fuel to be burned by an oxygen rich oxidant. These elements would be stored in tanks, and pumped into a reaction chamber, where the fuel would react. The resulting hot gases would be allowed to escape through a nozzle, thus providing thrust. He first understood and described the fundamentals of the relationships between fuelled mass, empty mass, final velocity and exhaust velocity, the energy of chemical reactions, and so on, as expressed by the Rocket Equation-he was thus the first person to put rocketry on a sound mathematical basis.

Realising that a single rocket would still not be able to attain the speed of 8 kilometres per second, Tsiolkovsky proposed that smaller rockets be lifted into space on the backs of larger ones-in what he called "rocket trains". This is the first known description of the "step rocket" with which all of our space activities to date have been conducted. In the step rocket a large rocket accelerates its payload to its maximum speed, and is jettisoned, leaving as its payload, a second, smaller, rocket travelling at high speed. This second rocket ignites its engines, and can add its final velocity to the final velocity of the discarded first stage. With modern high-energy fuels, light weight materials with strong heat-bearing capacities, the two-stage rocket is usually sufficient for low Earth orbital purposes. The Space Shuttle, for instance, jettisons the two solid fuel rocket boosters as its first stage, while its 3 high performance main engines power the vehicle on up to orbit. Since they burn right from launch pad to orbit insertion, the Space Shuttle could fairly be described as a "one and a half stage" vehicle. The two solid fuel boosters are recovered from the Atlantic, while the Orbiter and its main engines land, hopefully back at the launch site. Only the giant external fuel tank is thrown away, and proposals exist for pushing even that the remaining 1% of the way into orbit. By the time Main Engines Cut Off (MECO), and the E.T is dropped off at 9 minutes, the velocity of 8 kilometres per second has been attained, and the vehicle is in orbit. However, the 100 tons of Orbiter, with its 25-30 tons of Cargo, is but a small fraction of the launch mass of 3,000 tons-a mass ratio of 30:1. For a three stage rocket, like Ariane 4, which places up to 4.4 tons in Geostationary Orbit at 36,000 kilometres above the Earth, the mass ratio is even more dramatic-nearer 100:1, while Russia's two stage Soyuz launcher has a mass ratio of about 70:1.

Tsiolkovsky wrote extensively during the years 1883-1927, and considered that, initially, people would build space stations in orbit above the Earth assembled from the cargoes of fleets of "rocket trains". In the space stations, plants would provide most of the oxygen and remove exhaled carbon dioxide, as well as feeding the inhabitants; sunlight, and the raw , materials of the

Moon and asteroids would lay the foundations of a new and richer life in the vastness of space. From these stations, travellers would fuel up expeditions, and jump off to all the destinations of the Cosmos. These tasks would be greatly facilitated by a base just outside the main gravitational pull of the Earth. Tsiolkovsky imagined that the new environment would provide new industrial opportunities, and that peoples' health would improve, freed from the crushing gravity of Earth. He believed that advanced extraterrestrial civilizations are not to be found on some distant planet, but roaming free in gigantic "flying" nomadic settlements.

J.D.Bernal saw such large fabricated cities as possible interstellar arks in which Humanity would eventually migrate to other stars in epic voyages, lasting for generations.

Olaf Stapledon, in "First and Last Men", written in 1930, foresaw evolutionary changes in the nature of Humanity as he wandered far from the home planet. Tsiolkovsky put the date of the first artificial satellite at 2017, while Olaf Stapledon saw no interplanetary travel for over 100 million years.

The Founding Fathers.

Tsiolkovsky's ideas were not widely understood in his lifetime, but a group of enthusiasts in the 1920's in the new Soviet Union did see the possibilities, and formed a club named G.I.R.D to build experimental rockets. The 1920's were really the years in which practical rocketry began. The first liquid-fuelled rocket was launched by an eccentric "loner" in Spring 1926 in a bleak and desolate area of North Carolina. His name was Robert Hutchings Goddard, and his first rocket flew for 2.5 seconds at c.100 kilometres per hour. It reached the dizzy height of 56 metres, and produced no reaction except from indignant farmers as later models flew further and fell on their land! His attempts to persuade his countrymen that this was the road to Mars met with scant interest, but Theodore von

Karman of Caltech Jet Propulsion Laboratory began the interest of American industry in the infant fathered by Robert Goddard.

In Europe Professor Hermann Oberth-Transylvania's second most famous scion-began pioneering work on rockets in the mid-late 1920's achieving first flights of small rockets in the 1930's. He founded, along with Sanger, Winkel and others the Verein Für Raumschiffahrt for the development of experimental rockets. Initially, these were to be tested and used on racing cars for the Grand Prix, but the interplanetary motive was always there. The most notable recruit to the Oberth group was a young aristocratic engineer of 19 who was to make the major contribution to bringing the rocket from an enthusiast's plaything to the work-horse of space-Baron Wernher von Braun.

Oberth, too, had seen the possibilities in space-he produced early designs of the now classic doughnut shaped space station, in which rotation would provide a modest artificial gravity for the living quarters. His initial design was 50 metres in diameter, and would afford opportunities for astronomical observations, removed from the smoke and murkiness of Earth's atmosphere-this was the first step towards the Hubble Space Telescope, which, since its repairs in 1993 and 1997, is far exceeding the visionary's expectations. Oberth also imagined that large mirrors could be built to reflect sunlight into hitherto poorly lit regions of the Earth. The Russian Progress 20 unmanned cargo vessel demonstrated such a mirror in the early 1990's.

During the 1920's then, Russian, German, and American groups were laying the foundations of the rocket as we know it. Thousands of different chemical fuels were tested for "specific impulse" as well as methods of guidance, materials, and engine designs. These pioneers were largely enthusiasts working with their own time and resources, or those of wealthy patrons, or small engineering firms. By 1939 the elementary principles of rocket flight and guidance had been proven on the small scale; the efforts of GIRD, Goddard, Caltech, and the German groups had produced liquid-fuelled rockets which could reach 5-7 kilometres at several hundred kilometres per hour. The Germans were experimenting with the anti-aircraft Wasserfall rocket, while the Soviets had produced for the Red army an artillery rocket called Katyusha. Both these were to see service in World War 2, but were a long way from space travel. During the rush of enthusiasm in Germany in the early 1930's, the British Interplanetary Society was founded, which then, as now, produced valuable theoretical and research studies of space travel, rocketry, astronomy, and related topics. It was formed to promote knowledge, research, and enthusiasm in space related activities, and it is true that many space activists and professionals today-in America and Europe, and even in India and Israel-are members of the British Interplanetary Society and owe a debt of inspiration to it.

In 1939, a working party of the B.I.S, which included the young Arthur C.Clarke, published a study of a trip to the Moon, to be made by 3 men in a 1,000 ton three stage rocket. Many of the points of design were remarkably similar to the actual event in 1969, except, alas, the cost. The B.I.S project was priced at $1.4 million, but the rocketry was surprisingly prophetic. Apollo, in the event, cost $24,000 millions over 8 years-less than the cost of the cigarettes consumed in the U.S.A in the same period. The enormous complexity of the electronics and communications, again, was largely beyond the scope of the B.I.S. at the time, requiring revolutions in transistors, microcircuits, and computers. Even Arthur C.Clarke at that time never considered a flight to the Moon likely before the year 2000.

Then, as now, British officialdom showed a lack of even elementary understanding of the whole subject. Two examples prove the point. In 1939, just before Hitler's War, Arthur C.Clarke, on behalf of the B.I.S, wrote to the Air Ministry warning of possible major developments in long-range rocketry from the Nazis. The reply, which I understand is kept framed by Clarke at a secret location, states that His Majesty's Government "sees no future whatever in *Jet Propulsion*". Not much has changed. In 1987, another Clarke-Kenneth, 2 weeks before a meeting of the European space ministers, was asked his opinion on a British design for a revolutionary air breathing single stage to orbit spaceplane-H.O.T.O.L; he denied any Government interest in another supersonic Concorde.

It was in Germany, however, that the first true steps towards the space rocket were taken. In 1933 the Weimar Republic had been manoeuvred by the Nazi leader, Adolf Hitler, into signing its own death warrant, and with it Germany entered a new dark and militaristic age. At about the same time, the amateur rocket group of Oberth and his associates was running out of money, and work had slowed down.

Warfare and Spaceflight.

However, the young Wernher von Braun soon realized that the best bet for large-scale rocket development was to interest the military in the long-range artillery potential of large rockets. He succeeded in interesting a General Walter von Dornberger, who set up von Braun with facilities, and funds. It is not certain to what extent von Dornberger shared von Braun's interplanetary ambitions, but he was indispensable for von Braun's work. In 1938 von Braun and his team were moved to the top secret base of Peenemunde, a small island off the Baltic coast, and work began on the world's first long-range rocket. It was conceived as a 13 tons liquid fuelled rocket which would carry 1 ton of high explosive some 320 kilometres, thus putting London within reach. After 6 years, in the autumn of 1944, this rocket, named the V-2, began to bombard London, but, owing to Allied air raids, not in sufficient numbers to fulfil Hitler's expectations. In reaching an altitude of 100 kilometres, at 5,000 kilometres per hour, this vehicle deserves to be remembered as the first vehicle launched by humans into outer space; as so often in history, this step forward was born of the horrors of warfare. von Braun, at this time, was designing vehicles of greater range and power. One, the advanced V-2, had swept-back delta wings, and was intended to have a range of 640 kilometres, while the A-9/A-10 project consisted of a two stage rocket, which was to have carried 300 kilogrammes to New York. It is doubtful if these two projects were serious military projects. The military efficiency of hurling 300 kg. of explosive across the Atlantic at great expense with questionable accuracy would have been highly debatable. It is not entirely clear whether von Braun, in pursuit of an obsession, cynically made use of opportunities that became available at the Nordhausen prison factory, or if he was consciously supporting the aims of an evil regime. His unswerving ambition was always interplanetary flight, as his repeated upbraidings from the S.S administration made clear. The A-9/A-10 plans were always kept away from Himmler's prying gaze. The issue, as in nearly all such cases, is not whether the technology in itself is tainted with evil because of the unscrupulousness of one of its principal developers, but the use to which it is put. The fact of the *auto da fe* does not invalidate the use of fire. If, indeed, we can, over the coming decades, turn the military uses of von Braun's inventiveness to the task of opening up of the Space Frontier for all our descendants, this would surely be the best compensation for those lives that were brutally squandered in those dark days. For we can either turn our swords into ploughshares this way, or, by remaining Earthbound, into dust, along with their makers.

The requirements of British cryptography during the War led to great developments in another fledgling science vital to space technology-the electronic computer. The Enigma code-breaking machine, and the American Eniac, were the crude valve based forerunners of the computer revolution, and derived from the need to translate Nazi submarine, air craft, and spy coded messages. When it came to the true Space Age, these electronic developments were vital for the complicated guidance, telemetry, and navigation systems which became necessary. This was also true of the advances in radio, television, and radar, all of which also developed enormously during that decade 1940-50.

The key invention for the radios, electronics, and computers was the transistor, invented in 1948. This replaced the thermionic vacuum diode. The transistor does the same job, but is much smaller, produces less heat, and so is not liable to burn out or fuse. It also uses much less electrical power. These are essential for any component which has to work rapidly under extreme conditions, notably spaceflight. They are also the main element in the development of the Space Age which earlier writers could not foresee; they also account for the fact that actual space flight has turned out to be vastly more complex and expensive than the earliest pioneers had envisaged. However, further advances in materials and electronics will greatly reduce costs again, so that, in time, the pioneer writers may be nearer the mark after all.

In 1945, in Wireless World, Arthur C.Clarke published a most prophetic paper; he proposed that the emerging power of rocket technology be used to create a world-wide system of radio and telephonic communications. He suggested that the globe could be served by three relay stations placed at 36,000 kilometres above the equator at equal distances from each other, 120 apart. At this distance, an object orbits the Earth with a period of 24 hours, thus maintaining the same position over the Earth as it rotates. This orbit is called the Geosynchronous orbit or Clarke orbit. Clarke saw these stations as manned postal and relay stations, to be built, possibly before the century's end. What he could not be expected to foresee, was that, within a quarter of a century, such a global system would be in place, completely automatic , with the satellites weighing a mere few hundredweight each. The bigger ones weigh in at 1-3 tons nowadays. Indeed, the first comsats of the mid 1960's were little larger than a beer barrel, and considerably less subject to human interference! This illustrates the vital role of electronics in the birth of the Space Age, and the principle that seemingly unrelated developments, if brought together, can lead to revolutionary and unpredictable results.

After World War 2 came the Cold War between the two new superpowers-America and the Soviet Union-and the race to build the supernuclear H-Bomb. It was against this background that the V-2 rocket and the electronic marvels of the transistor came to be developed into the tools of the Space Age.

The United States managed to capture a small number of intact V-2 rockets after the War, and. for the first time, used the V-2 as the first stage of a two stage rocket. The second stage, a WAC Corporal sounding rocket, reached the then unheard of height of 400 kilometres. During this period, Chuck Yeager had proved that the sound barrier was no barrier at all, and that the way too higher velocities was open. The WAC Corporal / V-2 programme produced vital information on the temperature and pressure profiles of the upper atmosphere, and the outer orbital regions of outer space. These were essential in preparing re-entry capsules for missile nose-cones, and, eventually, crewed spaceships. The Soviet Union, meanwhile, had improved greatly on the V-2 design.By the early 1950's, the V-2 had been upgraded to the T-1 or "Victory" missile with a range of 880 kilometres. A launch facility was set up, in a village called Kapustin Yar, South East of Volgograd; this had been immortalized a few years earlier under its name of Stalingrad. From Kapustin Yar similar upper altitude research programmes began, with the use of dogs. In 1949 dogs were launched from Kapustin Yar and recovered from 400 kilometres altitude, thus demonstrating that living animals could survive the forces of a rocket launch, brief periods of low gravity, and a parachute assisted landing

At this time there was a growing spate of B movies and comics in the West devoted to the idea of space travel based on the new science of rocketry; there was a gathering awareness among writers and enthusiasts that here, at last, was a conceivable route to space. There was, however, no shortage of authorities to relegate such ideas to the "cereal box comic strips where they belonged". It was felt that humans could not survive the acceleration forces of lift-off-the heart or circulation would stop, or the astronaut would lose consciousness. Then, under zero gravity human circulation would stagnate, or he would go mad with disorientation. The vestibular system in the inner ears was programmed for gravity, and its disruption would be catastrophic. Then there were questions of life-support in a vacuum, difficulties of navigation, dangers from unknown cosmic radiation, and, looming far above all these hazards, the problem of returning a vehicle travelling at 8 or even 11 kilometres per second through the heat of re-entry. The use of engines to break the speed was out of the question; such a vehicle would be far too massive to be launched by any rocket under consideration. How was a vehicle to re-enter passively without being consumed by fire?

If this angle of re-entry was too steep, the vehicle would simply not lose enough speed, and would crash into the ground at high speed; if it was too shallow, the craft would skip back out of the atmosphere. The heating was likely to reach over 2,000 Centigrade, and melt any known metal; the crew would be cooked. In the early 1950's, many experts believed that short of the imaginary "Flying Saucer" with its unknown power source, dreams of human space travel were premature.

The "Space Race".

A deadly race between the two largest powers-America and the Soviet Union-was rapidly bringing to humanity the keys to space. For, after the spectacular and deadly proof of nuclear power at Hiroshima and Nagasaki, the physicists Oppenheimer and Teller in America, and Sakharov and Kurchatov in the U.S.S.R, realized that the energy of fission could be dwarfed by the thousand times greater power of thermonuclear fusion. A fission bomb core would be used to provide the trigger for the fusion of deuterium into helium-in explosive manner. This is a more explosive version of the fusion of hydrogen such as occurs in the stars, and could lead to a weapon of catastrophic power. In 1953 the U.S.S.R detonated the first "hydrogen" bomb, while America was not far behind. The logical carrier for such weapons to their enemies' camps was the new long-range rocketry being developed on both sides; the race for the intercontinental missile was on. Now the Russians had come first, but with a very crude and heavy "wet" bomb. In this bomb the deuterium is provided in the form of "heavy" water, and took a lot of room. This necessitated large boosters. By contrast the Americans had carried out a more compact solution; the deuterium was provided in the form of a solid blanket of lithium hydride, with deuterium in the place of the hydrogen. Ordinary hydrogen has a nucleus containing a solitary proton, whereas the nucleus of deuterium contains an additional neutron. The element Tritium has a hydrogen nucleus with two surplus neutrons. Deuterium is present in ordinary water in a concentration of 1 atom of deuterium for every 6,000 atoms of hydrogen.

The missile race also led to the solution of the problems of navigation and guidance, the development of telemetry for tracking of test warheads, and of materials for the problems of re-entry. Missile warheads were to travel at up to 16,000 kilometres per hour, and would experience considerable heating; not as great as for space vehicles, but the problem still had to be addressed, and tests provided essential information. Unexpected discoveries were made which were to be crucial.

All fast flying vehicles had tended towards ever sharper needle-point noses in order to cut their way through the air at ever higher speeds. If applied to re-entering warheads, the tips would rapidly become red-hot, and, in seconds, the whole nose-cone would be destroyed by heat. The first breakthrough came with the realization that a snub-nosed vehicle with a wide base was more satisfactory-the heat could be distributed over a wider area. If the vehicle were reversed upon re-entry, to enter with its broad base leading, heat could be deflected away from the sensitive parts of the nose-cone itself. This resulted in the curiously unstreamlined appearance of the Mercury, Gemini, and Apollo craft, which combine a modestly pointed cone for aerodynamic ascent through the Earth's atmosphere, with a broad base to distribute the heat of re-entry. Such considerations applied more recently in the "pepper-pot" appearance of the McDonnell-Douglas experimental DC-X "Delta Clipper" vehicle. The truncated cone shape of the first American space rocket was the totally unanticipated result of these experiments. Even this would not have sufficed, however, without the advances in materials science since the War. For ordinary metals could not survive the extreme conditions. What was required was a material which could be laid over the craft like an envelope which would soak up the heat of re-entry, without combustion, and with no conduction through to the body of the vehicle. On top of this, it also had to adhere reliably to the metal shell of the vehicle. Such materials as artificial rubber treated with epoxy resins were employed, and were known as Ablative heat shields. They had been used to line furnaces, and the nozzles of rocket engines before, but had to be vastly improved for the new stresses of re-entry. Thousands of compounds were tested, and, by 1955-6, a suitable solution was found. A miraculous material resulted which could be heated to 2,000° C on one side of a 2 cms thick sheet, while remaining lukewarm on the other side. It was also fireproof, and merely charred very slowly when exposed to heat. It was also light in weight, and could be glued to metal with epoxy resin glues. Further improvements were made to this ablative heat-shield material in the 1970's, for by that time these miraculous materials were required to perform flawlessly dozens of times, on the Space Shuttle.

Thus, by the mid 1950's all the pieces of the puzzle were in place-the long-range heavy rockets, the heat resistant materials for re-entry, the radio and electronic equipment for tracking and monitoring rockets during testing, and the computer technology for navigation and stabilising vehicles in flight. Also available was a range of suitable propellents as a result of years of testing every imaginable combination of chemical reagents, both liquid and solid. The supersonic aircraft research programmes of the X-series, and analogous trials in the U.S.S.R had yielded an understanding of atmospheric pressure and temperature conditions to the edge of space, as well as laying the foundations of life support systems in a vacuum. The aviation suits of the high altitude rocket-plane X series began to merge imperceptibly with the space-suits of the early astronauts and cosmonauts.

In 1957, the International Geophysical Year, it was widely expected that the United States would contribute by launching the first orbiting satellite. The U.S.Navy and Army were running separate rocket projects aimed at putting the U.S into the lead. The Navy's contender, Vanguard, was a small " minimal Launcher", which was expected to place into orbit a small package of instruments weighing a few pounds, whereas von Braun's group was pinning its hopes on the larger Jupiter-Redstone series. Inter-service rivalry and political indecision had led to many frustrations and delays; with sufficient backing, von Braun could have orbited a U.S satellite by 1955. The National Geographical Magazine of the period was looking forward to a U.S Earth satellite in 1958-9, amid great public scepticism. Many people, 300 years after Newton, wondered why such an object wouldn't fall down at once! In 1956, the Astronomer Royal described all this talk of men whizzing about the Universe in space suits as "utter bilge".

In 1957 the first Earth satellite was launched on October 4th, but not by the United States. The U.S.S.R team of Korolyov, Tikhonravov, Dushkin et al. had beaten them to it by launching Sputnik 1-not a mere grapefruit-sized object, but a massive 83 kgs. This created widespread concern in America that the Soviets had stolen a march in rocket technology, and hence in the critical nuclear arms race. Matters were not helped by the failure of America's attempts to follow suit; only succeeding finally in launching the miniscule 1.5 kilogrammes Explorer 1 satellite in January 1958. Meanwhile the Soviets had followed up with the half-ton Sputnik 2, in November 1957, with a canine passenger, Laika on board. Laika was put to sleep after a week in orbit. The U.S.S.R had achieved these feats directly as a result of her heavy I.C.B.M launcher programme; this had resulted in a rocket with approximately 400,000 kilograms of thrust, and which, with suitably upgraded upper stages, is still the main-stay of her crewed space programme. Today's A-2 launcher, which launches the Soyuz-TM and Progress cargo vehicles, is a direct descendant of the Sputnik launcher, with very few modifications. It has made well over 2000 launches, successfully, to date and is in production on an assembly-line basis.

By contrast, the Vanguard launcher was less than one quarter the size of the A-2, and, in the late 1950's, the Americans turned in desperation to von Braun. The U.S. in 1958 did begin to orbit satellites, and James van Allen made perhaps the most significant discovery of the first few years-the particle-trapping magnetic van Allen Belt which surrounds the Earth from 950 to 64,000 kilometres out. It is this belt which shields us from high energy cosmic particles from the Sun, and the galactic centre, which would otherwise have dire consequences for the evolution of Life on Earth. Thus, before the Space Age was a few months old, significant discoveries concerning the ecology of the Earth and upper atmospheric conditions were already resulting. President Eisenhower had seen little point in being concerned because the Russians were able to launch a "tin grapefruit" above the Earth, and was ill disposed to any American attempt to follow suit.

The Soviets, however, continued to make the running. During 1958-60 their Sputniks increased to 4.5 tons, and live animals were recovered in soft landings in Kazakhstan. Also, launch facilities were set up to the North of a rather dilapidated walled mediaeval town called Tyuratam. Railroad branches, an airport, and a power station were installed in this area, and, over the next few years, the village of Tyuratam swelled to 50-60,000 and became a powerhouse of industry in the steppes.

Also in 1959, the Soviet probe Luna 3 transmitted to Earth the first ever views of the Lunar farside. By 1961 the Soviets were ready to take the next giant step and orbited the world's first spaceman, Major Yuri Gagarin, aged 28. He orbited the Earth once, in 89 minutes on board the spaceship Vostok 1. This craft weighed 4.5 tons, and consisted of a spherical cabin plus re-entry module, about 3 metres in diameter, with a cylindrical service module, containing retrorockets, air and water supplies, and electrical storage batteries. This latter was detached from the re-entry vehicle after the craft was placed in its re-entry corridor, and the sphere was landed in Central Asia by parachute braking, with retro-rockets during the last few metres of descent. In August 1961 their design was well proved, and Gherman Titov in Vostok 2 had orbited the Earth 17 times, in 25 hours. Thus, by 1961, human space flight, a dream for so many centuries, had become an established fact. Talk began to be heard of Soviet trips to the Moon, space stations, and the ideological superiority of planned Communist societies over decadent American capitalism;

Nikita Khrushchev was not slow to point to the space successes of the U.S.S.R as evidence of her superiority as a model for Third World countries. In that period, the world was seen very much as a chessboard on which the Americans and Soviets were playing for very high stakes. Apart from ideology and propaganda, anxiety was expressed about the dangers of orbital espionage, and, later, the possible emplacement of nuclear weapons in space. It was Russian artillerymen who had helped to win Stalingrad and Kursk for the Red Army, and it was classical military doctrine that the first strategic objective was to place your gunners and spotters on higher ground overlooking the enemy. No doubt that is how Soviet military strategists saw matters-it was the hard-headed Marshal Stalin who initiated the Soviet space programme, not the flamboyant, propagandist, debating Nikita Khrushchev. The latter, however, showed a keen appreciation of the showman value of spaceflight, and loved to be seen embracing the first cosmonauts. This continued under Khrushchev's successors, Brezhnev and Kosygin, with the result that the Soviet efforts became distorted by the need for an ever increasing succession of "space spectaculars", which led to many false starts, and did little to advance the real needs of space development. The death of Korolyev removed a powerful guiding hand from the programme, so that it came to lack focus, and fell foul of political rivalries. Despite hot denials at the time, there was a Soviet programme to beat the Americans to the Moon, but it foundered for the reasons given above, plus the deplorable state of Soviet electronics and computers.

The new President Kennedy saw himself facing an assertive and combative U.S.S.R, which was trying on a global scale to find his mettle-Fidel Castro had recently gained power in Cuba, Khrushchev was threatening over Berlin, and, on top of all this, Russia appeared to be stealing the lead in advanced technology. After Gagarin, and Titov, and the Bay of Pigs, Kennedy badly needed a morale booster, and a technological stimulus for America. So it was that, in 1961, Kennedy entrusted to the newly created civilian space agency, N.A.S.A , the task of landing an American on the Moon and returning him safely to Earth by 1970. Von Braun's time had come at last, and the gigantic rockets could begin to leave his drawing-board.

This was a great challenge to a nation which had not yet even sent a man into space, but promised a clear prospect of demonstrating American technological superiority over the Soviets. It gave them a clear measurable goal, and, in their effort to fulfil it, led to striking industrial, medical, and scientific developments and spin-offs. It also appealed to the pioneering and Frontier spirit in the American psyche. In 1961 it was by no means clear among astronomers that the Moon had a solid as opposed to a dusty surface, so that a landing was by no means a foregone conclusion.

The development of large rockets during the 1961-7 period led to, firstly, a new scale of business and industrial organizations. The sub-contracting of thousands of specialized small parts manufacture-fuels, engines, electronics, guidance, life-support systems and so on-among many firms over a whole continent led to new heights of precision engineering and co-ordination to the nearest thousandth of a millimetre, in some cases, between products manufactured at sites separated by hundreds of kilometres. The new sophistication in management and technical skills implied in this was one of the first, and longest lived, benefits of the space programmes of the U.S.A. and the U.S.S.R. More recently, this has been a vital contribution to the economies of newer space powers such as Europe, China, India, Brazil, and Israel. This new expertise began to spill over into other areas of the economy, and the development of a "high tech", post industrial, sophisticated economy began to take shape at this time. Considerable employment in service and supporting industries also resulted from the space programme; it is estimated that the Apollo programme created and sustained 2 million jobs of varying degrees of skill by 1969-70, and that the run-down in space activity cost many hundreds of thousands of jobs, with the concomitant recession. In 1992, two eminent economists submitted evidence to the United States House of Representatives that space activities are one of the greatest known Keynesian economic multipliers-this with a Shuttle programme limping along in the wake of the Challenger disaster of 1986. For every U.S dollar spent in space, 7 more are earned on the Earth. The Soviet Union and, since then, other powers have seen in space a potent "driver" for technically backward economies. Traditionally, this role has fallen to warfare, as in the World Wars. The Second World War, in particular, is widely credited with having ended the 1930's recession. War is now too costly and uncontrollable in human and economic terms for this or any other purpose-indeed, we have now reached the point when advanced weapon systems are so ruinously expensive that armaments industries can only survive by selling to their owners' enemies. The recent Iran and Iraq arms sales scandals in the UK prove the point decisively. This is the meaning of Arthur C. Clarke's well known description of space exploration as a "moral equivalent of war"; he also cited space rivalry as a sublimation of national rivalries in the early 1955-70 Space Age. In this first phase of exploration, culminating in the Apollo landings, the motives were national prestige, propaganda, and security first, exploration second, and direct terrestrial benefits third.

Many thought that after the Apollo fire in January 1967, and the Soyuz 1 tragedy in April 1967, the Soviets had given up racing for the Moon, thinking that the USA could not achieve success; however the attempts to develop the gigantic N-1 Moon rocket are now known to have been parts of a serious effort to reach the Moon by the Soviets, while the Proton rocket was to have launched a heavy Soyuz carrying 2 cosmonauts and equipment to dock with a lunar transit lander and return complex to be launched into earth orbit by a giant N-1. This was more powerful even than the Saturn V. Thoughts of a permanent Soviet lunar base were even voiced as Vladimir Barmin, Director of the programme revealed in the recent collection of interviews entitled "Roads to Space" published by McGraw-Hill and advertised in their Aviation Week Newsletter. Of course continuing difficulties in co-ordinating the 32 engines of the N-1's first stage, exacerbated by the death of Sergei Korolev and the rivalries between Soviet space industrial contractors, led to the US triumphs; nothing concrete was heard of Soviet lunar ambitions for over 20 years. Despite the conventional view of the 1970's and 80's there really was a Space Race to the Moon, and the USA won it.

We have seen in this chapter the technological steps and political factors which led to humanity's first steps into space, and how they took place far sooner than expected. The crucial question to be answered next was, how does the human organism itself react to the new environment opened up by our technological prowess? In the next chapter we shall explore the human adaptation to space, and see that, although far-reaching changes occur, the body adapts far more successfully than had been thought possible, and that future developments offer clear possibilities of meeting any long term goals adopted by Humanity in space.

Chapter 5

Man, Medicine, and Space.

We have seen how, in the early 1960's, America and Russia arrived much sooner than anybody had expected at the situation where crewed space flights could be undertaken. There were, however, very real questions about the ability of the delicate human organism to withstand the rigours of launching, and then the strange environment of greatly reduced gravity.

During the Vostok flights, of up to 5 days' duration, and the Mercury and Gemini missions of up to 14 days in 1966, the main physiological problems of space travel for short flights were explored, understood, and coped with. The fear that human beings could not survive the stresses of lift-off was rapidly disposed of. The early rockets reached a maximum acceleration of over 4 G, for a mere 3-4 minutes, during which there was at most a temporary swooning, which rapidly subsided. The Space Shuttle today lifts off more slowly than the early rockets, with a maximum of 3 G acceleration, which is well within the tolerance of most fit adults. In recent years, men in their late fifties have coped with this perfectly well. There remained the issue of zero g and human adaptation to this novel feature. Gemini flights worked on the principle of approximately doubling time for flights; from 1.5 days of the last Mercury mission, to 4 days for Gemini 4, 8 for Gemini 5, and 14 for Gemini 7. The lunar landing flights were expected to last for 10-11 days, so that the first requirement was to cope with 2 weeks in space. The later Gemini flights centred on the technical requirements of rendezvous and docking, an area in which the Soviets conspicuously lagged behind, until the late 1970's, with the advent of modified Soyuz craft fitted with better electronics, and the Salyut 6 and 7 space stations.

The short term effects were less severe than many had expected, and centred on three main areas, which have, in the later Skylab flights of up to 3 months, and the Salyut and Mir missions of recent years been extensively studied. Over the period 1978-1986, The Soviets pushed the endurance record for human space flights from 3 to 7 months, while on Mir, the multimodular space station begun in 1986, and continously crewed for 10 years until evacuated in August 1999, this duration has been pushed to 14 months by a physician, Dr.Valery Polyakov. Yuri Romanenko spent 4.5 months in 1978 on Salyut 7, and then 10 months on Mir in 1986/7. He was observed to be fitter after the latter flight than the former, and was able to appear on TV the day after landing. No real restrictions have been found in human productivity in flights of up to 6 months duration so far. From the missions of Romanenko we can confidently expect human adaptation to space for flights of up to 12 months, and probably longer, without more than 2-3 weeks readaptation to Earth's gravity. Titov and Manarov spent the year 1988 on board Mir, thus establishing the 12 month range nicely. The return of Dr Polyakov in March 1995 without serious ill effects, and of Dr Norman Thagard after the historic Mir/Atlantis docking in June 1995, will provide data on the long-term effects of microgravity, while later American and European missions to Mir of up to 6 months duration has greatly widened the data base. Results from Dr.Thagard's mission showed that his problems were more cultural , and psychological. Despite being advised to leave the Shuttle in recumbent chairs because of cardiac de-compensation, Thagard was able to stand upright for 10 minutes shortly after landing, without ill effects. Much valuable input Dr. Thagard, and the following 7 preliminary Russian / American flights will find its way into the managerial aspects of the International Space Station. The female cosmonaut Ms.Yelena Kondakova, also made a landmark flight of 6 months in 1994/5, so that both sexes will have contributed to this important work. As for the longer term prospects of longer stays in space, we shall consider how these are affected by present medical knowledge, and how they will be solved in the coming generation.

The most serious problem from the short-term point of view is the "space sickness syndrome". This arises from the fact that in a space ship there is no up or down. On Earth, humans and other animals balance themselves by reference to three main systems; the eyesight unconsciously relays to the hindbrain and cerebellum information concerning orientation; observation of the lie of objects such as trees, the floor, and so on can be compared with memories of where such things ought to be, and a sense of up and down can be derived. This can be seen in the aqualung swimmer who, in clear water 15 metres down, may not see the bottom. He can, however, find the surface by seeing that "upwards" contains more diffuse sunlight in contrast to the inky depths, and that his exhaled air bubbles rise towards the surface.

The sense of touch, measures the pressure of the ground on the soles of the feet, and is integrated via the hindbrain to give a sense of whether the ground under one's feet is above or below the head.

Finally, the third system is the vestibular, or sense of balance. There are, inside the inner ear, 3 semicircular canals orientated to each other at right angles. These contain fluid, with small stones or otoliths, free to move inside them. Acceleration in any direction leaves the stones standing, trying to catch up with the movement of the canals themselves. This happens in seasickness, and the results are familiar to all of us; the vestibular nerves register the excessive movement of fluid in the semicircular canals, and activate the vomiting centre of the brain. In space there are similar disturbances owing to the lack of perceived up and down. However, the "space sickness syndrome" has its differences; it is more usually a feeling of lassitude, or feeling below par, rather than outright nausea and vomiting as in sea sickness or car sickness, in those subjects who are prone to it. There is some correlation between proneness to sea and space sickness, but crew selection to avoid it is far from simple. According to Helen Sharman, the

Russians do use susceptibility to motion sickness as a partial screen, although non-sufferers from terrestrial motion sickness cannot be guaranteed a smooth ride in space. Space sickness affects about 40 % of space travellers, is of short duration, lasting 1-5 days, with an average of 2 days. The underlying mechanisms are complex, and still not fully understood. It differs from the terrestrial motion sickness in the ways described, and also in the fact that, in a boat, the semicircular canals are being violently shaken about in various directions, with the sickness resulting from the disparity in the motion of the little stones. In space, there is a relative *lack* of stimulation of the semicircular canals, owing to the absence of gravity. Recently, Canadian space physicians have suggested a pre-emptive treatment which can prevent or greatly mitigate space sickness. They report that, if a person reads for half an hour in a moving motor vehicle each day for a week, conditioning of the vestibular system can result in excellent results in space. Better to be queasy on earth, in a car, than up in space. Pharmaceutical products such as prochlorperazine or cinnarizine are effective, but, because of their potential sedative effects, are not suitable for professional astronauts, who may need their wits about them. For tourists, it would be another matter entirely.

Adapting to Space.

There is a separate period of adaptation and space sickness on moving from one space vehicle to another, especially if the second one is larger. This was seen in crew transfers between Soyuz and Salyut or Mir, and Apollo and Skylab, during the early space station experiments at the end of the first phase of crewed space flight. If this was simply a problem of semicircular canal adjustment to microgravity, it is hard to see why transfer from one microgravity environment to another should bring about another bout of the syndrome.

Space sickness has been a troublesome problem on short flights and limited their efficiency, but has not proved disabling, and there is no evidence of any long term effect on human physiology. Indeed, it has not prevented complex operations by the crews of both programmes on several occasions. Space sickness results not so much from the violent disturbance of the semicircular canals as their failure to work at all! The body suddenly has to orientate itself without their assistance, and must largely fall back on sight and proprioception. A sense of mental and physical exhaustion more likely results from increased strain on these less used senses-rather as "eyestrain" is believed to result from overuse of the eyes in poor lighting. The body forms a frame of reference from unconsciously noting the relative positions of objects in the craft, and "programs" an arbitrary standard of up and down into the memory. On transfer to another, especially a larger vessel, the brain must "reprogram" itself. More subtle disturbances in perception will be discussed a little later-these seem to add credibility to my suggestion.

The early Mercury and Gemini space flights took place in profound "ignorance"; the effects of weightlessness on the heart rate, blood pressure, respiration and fluid balance had all to be investigated. The early spacemen needed to be continually monitored, and the telemetry and sensor equipment needed for medical experts at Houston and Tyuratam had to be developed from scratch and miniaturized. Since the windows of the Mercury spacecraft were less than 15 centimetres from the astronaut's face, there was not much room for equipment! The miniaturized sensors, electronics, and transmitting devices soon paid dividends in emergency units, cardiac care units, and medical technology the world over; the crewed space programme became a pace-setter in this area.

The requirements of space suits, and extravehicular activity, led to the development of lightweight heat retaining aluminium foil materials, backed with latex. Examples are thermal insoles, originally developed for lining astronauts' boots, but very effective in stopping the cold from street pavements striking through to one's feet, and also a 75 gram folding blanket which folds away into a cigarette packet-this can keep two people alive for up to two days in harsh exterior conditions-a life-saving part of mountaineering or rally driving equipment.

The requirements for minimum food bulk, and processing into paste for use by astronauts, led to a search for improved methods of preservation and storage. A new trilaminate plastic was developed for use by N.A.S.A in which many appetizing dishes could be initially boiled and packed, and then have a shelf life of many years. Many years later, the meal could be eaten straight out of the packet, or boiled for 7 minutes first, or microwaved. Tests showed no deterioration in vitamin or nutritional content after three years-long enough for a return trip to Mars-or Antarctica. These packs cost between $1 and $1-40 each, and have been satisfactory on desert and Polar expeditions. Their makers in England have been exploring their possible use for pensioners or folk otherwise disabled.

We have seen the emergence of tele-medicine in the Armenia earthquake of 1978, and in the NASA run programme in 19 South American states, which allows, via satellite communications, consultations between field workers at remote or disaster-stricken locations with the best available specialists. This application has self-evident applications throughout the world, from Africa to Central Asia, and saves lives down on Earth. The University of Surrey's Healthsat programme has been helping midwives in the African bush for years now, with excellent results and at low cost.

The early Space Age, and the missile programmes, forced the process of miniaturization, and the introduction of third and fourth generation of microprocessors and computers. The present silicon chip revolution is in some measure a child of the larger space revolution now in progress; it is also feeding back into the space movement, and further advances will come from future activities in space. The advent of virtual reality, and the drive towards "cheaper, smaller, faster" spacecraft under NASA administrator Dan Goldin's leadership will bring space imagery increasingly within the purview of the education and entertainment industries. We shall see space-borne astronomy, lunar, and possibly Martian sight-seeing come within the ambit of private enterprise and educational establishments.

The widespread use of non-stick teflon materials coincided with the early space programmes. This resulted from the need to line rocket fuel feed lines and pumps so that maximum use of fuel could be ensured, and that no residue would cause a risk of explosion.

As space flights became longer, with Soyuz 9 lasting 17 days, Skylabs 2, 3 and 4 lasting 28, 59 and 84 days respectively, and later Salyut and Mir missions endured for up to 437 days (Mar 1995), other physiological adaptations of both sexes to space flight were explored. The cardiovascular system was soon shown to be markedly affected by microgravity. In the first few hours, it was discovered that 1-2 litres of bodily fluid, normally distributed in the tissues of the legs, were rapidly re-distributed into the upper body and larger thoracic veins. Initially, the amount of blood draining into the heart increased, and that this led to an increased loss through the kidneys. The result of this is a loss of salt and water, followed by increased thirst and drinking. Subjectively, in the early stages of this re-distribution, there is a feeling of fullness in the head, rather like a mild sinus infection, without the pain.

This was a totally unexpected finding, which will lead to understanding of the way in which water and salts are distributed and controlled within the body. Classically it has always been thought that the amount of fluid in the tissues of the body, outside the small blood vessels called capillaries, depends on just two forces. At the start of the capillary network, the arterial end, this pressure is about 32 millimetres of mercury, while at the venous, or draining end, it has fallen to less than 10. There is, therefore, a net pressure gradient of about 25 millimetres of mercury between the arterial and venous capillary circulation which tends to push fluid out into the surrounding tissues. Without a counterbalancing "pull" inwards the peripheral tissues would simply fill up with fluid. The legs do this in various medical conditions, such as heart failure. This results partly from an increased pressure in the venous end tending to lead to a build up of fluid waiting to be pumped around the body by a failing heart. If the heart pumps less effectively, less blood per second reaches the kidney, and the kidneys are less able to maintain fluid balance.

The opposing pressure is "colloid osmotic pressure". In school experiments, if water containing different concentrations of a salt are placed either side of a membrane with very fine pores-a semi-permeable membrane-water will be "pulled" from the more dilute to the more concentrated solution by an osmotic pressure until the number of particles on each side of the membrane is equalized. If there is a mixture of dissolved molecules, the total number in each side of the membrane is equalized. In the capillaries, large molecules called proteins-specifically the albumins-act as the osmotic "puller" of fluid from the spaces outside the vessels. These albumin molecules are too large to leak out of the capillaries, unless the lining cells are damaged by poisons, infection, or wounds. In these situations one often sees local tissue swelling for this reason; this observation is familiar to most people. The osmotic pressure exerted by blood proteins is about 25 millimetres of mercury; this suffices in health to allow fluid and nutrients to be pushed out into the tissues from the arterial end of the capillary networks and fluid loaded with waste products to be pulled in at the venous end. In diseases there are many ways in which this happy balance can be upset. We have mentioned heart failure, and local infection. In liver disease there is a general reduction of manufacture of blood albumins, so, that there is a reduced osmotic pressure; it is not therefore surprising to find evidence of tissue distension by fluid, in severe cases of liver disease.

It has always been assumed that this balance of hydrostatic and osmotic pressures, coupled with the integrity of the capillary wall linings, was the sole determinant of fluid distribution in the peripheral tissues of the body; the force of gravity imposes a gradient between the heart and the feet, and that its absence forces a new adaptation. This has been extensively studied in the Life Sciences missions of the European Spacelab, and, of course, on Salyut and Mir.

Human Metamorphosis

The heart and circulation adapt over the longer term to microgravity; over the first 6 weeks of a long-stay mission, the output of the heart diminishes to one third of normal, the pulse rate and blood pressure become lower, and the blood cholesterol level falls to up to one third of pre-flight levels. The total volume of blood diminishes, as the output of the kidneys increases at first in response to the accumulation of fluid in the chest. In some of the Skylab astronauts, disturbances in the rhythm of the heart were noted, probably due to a combination of "deconditioning" and altered blood chemistry. A self-regulating, electronic pace-maker girdle was developed to be worn around the astronauts' chests, which would automatically detect alterations in rhythms and correct them. This device is likely to find its way into several medical applications, notably those in whom a surgically implanted pace-maker is impracticable. Also in this area externally rechargeable pacemaker batteries of light compact construction and long-lived reliable operation have been derived from the requirements of unmanned space satellites.

This cardiac "de-compensation" reaches a plateau after 6 weeks, at which time a new physiological balance has been struck, and further deterioration does not take place. Dozens of Russians and Americans have now flown for well over 6 weeks, including one woman, and men well into their forties, so that a good sample of subjects is now available. The Russians have logged several very long missions-of 7, 10, 12, and 14 months duration-while 5-6 months is a routine tour of duty on Mir, which was almost continually crewed from February 1986 until August 1999. Crews have reported that they have performed their tasks-often including long and arduous space walks and repair jobs-efficiently in space, even after long periods of microgravity. Current practice is to prepare the heart and circulation for re-entry into Earth's gravity by the use of lower limb pressure suits to tone up the leg muscles, and 2-3 weeks of extra fluid and salt intake to top up the circulation, and recondition the heart.

The main cardiac problem for spacemen to date has been the need for sudden readjustment to 1 G on return to Earth-a process taking at most 3 days. Rigorous programmes of exercise of 1.5 to 2 hours per day have been employed on Skylab, Salyut and Mir in an attempt to reduce de-compensation in space. On one of the Salyut flights, the Cuban guest cosmonaut, Arnaldo Mendez, experimented with a pair of boots which acted as an automatic pump to speed the circulation, as well as exercising the leg muscles.

On landing, cosmonauts usually report mild problems of breathlessness and weakness, and so are whisked off to a hotel for 2-3 days of debriefing and recuperation. The sudden assumption of an upright posture can produce fainting-which is avoided by carrying the cosmonauts away from the landing zone in recumbent seats. The evidence to date is that these problems are mild, and of brief duration. In Communist times, they usually had to face the media, and a compulsory embrace from the General Secretary, within a week of landing, and have always coped very well! Russian experts have pronounced that the cardiac issue in space flight had been mastered, and that the way to Mars was now open, from a bio-medical viewpoint. Their decision to proceed to a 14 month flight shows a degree of confidence.

Astronauts and cosmonauts are often in their middle forties, and rising beyond 50 in some cases; although fit specimens, at this age they are hardly likely to be supermen from the cardiovascular point of view. The flight of Alexandrov and Lyakhov in Salyut 7 provides excellent evidence of human ability to work efficiently in space. Following a successful 3 months of experiments on board Salyut 7, it was decided to send up an additional 2 cosmonauts in September 1983 to repair a fuel leak in the Salyut's attitude control system, and to fix new solar power panels to the station. However, their A-2 launcher developed a fire on the launch pad, and the new crew had to abort the launch. The Soviets debriefed this crew, and sent up the unmanned "Progress 18" tanker/ supply vessel to the Salyut, which bore spare parts for mending the fuel leak, as well as the two solar panels. In two marathon space walks of 3.5 hours each, the two men successfully fitted the two large panels-a heavy and complicated operation. This job was evidently meant to be carried out by a fresh crew, or by two crews, making four men. In any event, Lyakhov and Alexandrov performed the task well, and were fit enough shortly afterwards to appear on Moscow TV within days of ending a 5 month mission. They admitted to no more than tiredness. Such repairs, incidentally, and that performed on Skylab in 1973, as well as the repairs of the Solar Max satellite in 1984, and the Hubble Space Telescope in 1993 and 1997, and many others over the past 15-20 years, are eloquent proof of the justification for a human presence in space. Some s pace hardware is now too sophisticated, expensive, and important to be allowed to fail simply for want of a human repair capability. Solar Max was providing vital evidence of the variability of solar output during its 11 year cycle, when it broke down, necessitating a repair mission. Without repair, the crucial series of data during the whole of a solar cycle could not be obtained, and much information would have been lost. The billion dollar Skylab in 1973 suffered loss of a solar power panel, as well as a protective temperature regulating panel, on launch; without an improvized repair by the first crew, the whole venture, together with its invaluable bio-medical information and 50,000 images of the Sun in various wave-lengths, would have been lost

The third major physical result of prolonged space flight proved to be a reduction in blood and bone calcium, with increased loss in the urine. Studies on Skylab 2 showed that the rare of loss was about 0.4 to 1% of total body calcium per month of microgravity. Over long flights, this led to reduced bone mass, and there is no sign to date that the rate of loss eases up. It might well be that in very lengthy flights of 2 years or more, with no gravity or artificial equivalent the bones could atrophy, and the human body become invertebrate. The bone softening has caused no problems in flight so far but could, in principle, lead to bone pain or fracture. This effect is not diminished by dietary calcium, but is abated by fluoride dietary supplements, as well as rigorous exercise programmes. Certainly, the loss of body weight, which is a good index of the loss of bone and muscle mass in spaceflight, had markedly diminished by the time of Romanenko's second record-breaking mission-indeed, he actually gained a little weight! Laboratory experiments have already shown that fragments of cultured bone grow or heal in response to applied electro-mechanical stress, while prolonged bedrest has been known for a long time to lead to a similar softening of the bones, so that, on reflection, these findings are less than astonishing.

The study of the problems of bone calcium loss in microgravity is emerging as a possible model for the widespread condition of osteoporosis, and it is hoped that the International Space Station will make useful contributions in this area-both in understanding the condition and in evaluating possible treatments. It may also well turn out that some variant of the currently used Hormone Replacement Therapy, now widely offered to women of menopausal age to prevent later "brittle bones", may have a part to play in the prevention of this problem. Even after 2 years of microgravity, the calcium loss on the foregoing calculations will amount to 10-15 % or so of the total body calcium; it seems unlikely that the serious problems of fracture will occur even at this stage. In any event, regular exercise stints on board such craft will greatly reduce this hazard, as has already been shown. The introduction of the "Chibis" suit, in the later Salyut missions, has greatly improved the efficacy of the exercise regimes, and, following the Romanenko mission of 12 months, Russian medical experts pronounced the problem "solved". The "Chibis" suit is a kind of close-fitting elastic pair of leggings, with creates artificial resistance to leg movements, thus demanding more effort, and generating exercise.

Associated with these bone losses is a wasting of muscle mass; about 70% of muscle mass exists to overcome gravity here on Earth. In space this becomes redundant, and the larger muscle blocks diminish in bulk. There are two sorts of muscle fibres; rapid firing "red" muscle fibres as used in sprinting, and more slowly acting "white" muscle fibres, as used in isometric exercises or sustained effort. The evidence to date suggests that there is no advantage, even a slight disadvantage in super-athletes' adaptation to microgravity. The red fibres fare worse in space, as there is very little scope for sprinting in a space station, and exercise confers no benefit in space stations to the athlete over a reasonably fit scientist. This is heartening news for enthusiasts and would-be tourists and colonists, many of whose devotion to exercise may be less than whole-hearted! Again, in flights of up to 3 months, muscle wasting is scarcely a problem, while in the Russian long-duration missions to date, the crew show an unsteadiness on their feet which recovers within 2-3 days of landing. Some crew members have required assistance on leaving their Soyuz capsules, and one or two days in a hotel to recover their "land legs". It would appear analogous to the similar problem which greets long duration sailors, who have likewise to regain their land legs after long around the world voyages. Vigorous programmes of exercise both active and passive, are of proven value here, and Mendez' Cuban boots have proved their worth, along with the Chibis suit.

Flights of up to 6 months are now proved to be well within the physiological limits of fit men of middle age. The conditions are also extreme-confined space in zero gravity. The Russians are clearly trying, patiently and methodically, to find the limits in extreme conditions, and, even after a 14 month mission, have not reached them. Working space stations in the future, such as the International Space Station , as with the Mir complex, are and will be manned on a 5-6 monthly shift system, which has been shown by Russian practice to be well within the range of good physiology, and to pose no serious problems of re-adaptation on the ground.

As we enter the age of longer duration missions, whether for prolonged interplanetary voyages, or for large-scale manufacturing and construction tasks, space vehicles will become larger, more like habitats, and the gravity problem will be resolved. Experiments on the Cosmos series of unmanned satellites have shown that rotating a spacecraft around a fixed point provides a centrifugal force which substitutes for gravity, and greatly reduces the physiological effects of microgravity. In large space stations, interplanetary vehicles, or manufacturing colonies, artificial gravity can be provided by the simple expedient of rotation; on a trip to Mars, the gravity can be gradually altered during the 8-9 month passage so as to prepare for life on the red planet. It is, of course, quite possible that the use of nuclear or electric-ion propulsion systems will be used to reduce the cruising time to Mars to something more congenial, like 6-8 weeks, which would avoid this issue entirely. A large size is required to provide reasonable centrifugal force without giddiness, but stations to be crewed for 1 or 2 years or more will of necessity be large. It can be said in summary that settlers and workers of the future need never be exposed to more than three months of zero gravity at a stretch, unless this is desired. Similarly, modern airline passengers are not exposed to the unpressurized cold endured by Alcock and Brown in 1919. We travel in comfort, and so will our descendants on their voyages to distant worlds.

At the Second Annual Space Development Conference, in Houston, of the then L-5 Society, astronaut Gerald Carr of Skylab 4 summarizes the position with the benefit of 84 days flight experience as follows. for the first 4-6 weeks, there is adaptation of orientation and perception, as well as a general weakening of cardiac, bone, and muscle mass; after 6 weeks his crew felt at home, and could work normally, and in many respects could perform tasks with greater speed and productivity than on Earth. Man is an adaptable organism, and performs remarkably well in the new environment. Other subtle changes in perception were described. For instance, on Earth, a thrown object describes a curve, or arc, in its journey through the air. In space, by contrast, the object would fly in a geometrically straight line. It takes several weeks for the brain to learn the "new mechanics" of space, and also to get used to leaving objects parked in midair during laboratory work!

Other effects noted on Soyuz 9 and Skylab flights were a decreased activity of blood lymphocytes-these are cells which make antibodies, and are important in a class of diseases called "autoimmune" diseases. A typical, and all too common example, is believed to be rheumatoid arthritis. Current theory has it that an as yet unknown foreign body, probably a virus, invades the body and produces, naturally, an immune reaction.

This reaction is somewhat over-zealous and partially misdirected, so that the "police cells", or lymphocytes, attack the cells of joint lining tissues as well as the hypothetical antigen. The resulting fracas causes joint tissue damage, known as arthritis. This is a very simplified picture, but forms a rough description of a number of painful and debilitating conditions. The implication of this reduction in lymphocyte activity is a reduced resistance to infections; a visiting crew who were ill with colds might produce an epidemic in a space station, or colony. Visiting crews to space stations or settlements will need to be carefully screened for infectious diseases, while personnel returning to Earth on vacation after long stints in settlements may well need booster immunizations, or passive immunoglobulins, to some Earthly diseases. However, prolonged occupation of a space settlement might well lower susceptibility to auto-immune diseases. The Space Shuttle mission, STS 10, which lost two communications satellites-retrieved and launched by a later mission-demonstrated that arthritis in space was rapidly alleviated by microgravity. There is also a microgravity induced anaemia, again probably resulting from reduced metabolic demands on a body freed from the bonds of gravity.

A God's Eye View?

Somewhat surprisingly, during the rest periods when crews looked out in the wonderful panorama of the Earth revolving "beneath" them the eyesight became much keener, and the eyes were able to resolve much finer detail on the Earth than had been calculated as possible-the vapour trails of aircraft, and wakes of ships, for instance, could be made out much more clearly. If my earlier suggestion that the body reacts to the deprivation of the sense of balance in the inner ear, and lack of proprioception, is valid, then the compensation for this deficiency by improved, sharper, eyesight, makes a lot of sense. Just as many blind people report a degree of compensation in their other senses, especially hearing and touch, so the gain in eyesight of astronauts seems a natural conclusion. It appears, therefore, that one of the benefits of microgravity is a God's eye view of the Earth and Heavens.

Another issue that will become important during the next phase, of interplanetary travel, will be that of radiation. This comes to us in the form of charged particles, atomic nuclei-cosmic rays-or gamma radiation. The chief source of charged particles-electrons and alpha rays-is the Sun, whose output is very variable. For the most part, spacecraft keep out such radiation but, in a solar eruption, or flare, the Sun, with barely 24 hours warning, can put out lethal doses of high energy protons and electrons. On entering the Earth's upper atmosphere, these strip off electrons from the rarified atoms in the ionosphere, resulting in the Aurora Borealis, or Australis. These can be stopped by thick shielding, of any material. Since interplanetary missions must needs take large quantities of fuel and water with them, suitable design of spaceships can use these as bulkheads between the Sun and the crew. The bulk of the radiation hazard is due to particulate radiation from solar flares, which at present cannot be predicted with sufficient accuracy.

Missions like SOHO (1996) and TRACE are improving understanding of solar magnetohydrodynamics, and, in time, should lead to better flare prediction and assessment. This radiation provides the biggest threat, since the background solar radiation and the galactic cosmic radiation is consistent, well known, and, on a trip to Mars, gives a relatively small increase in cancer risks to traveller. The dose on a 3 year round trip to Mars is approximately equal to a lifetime's exposure on Earth's surface, and is considered acceptable.

Long-standing structures can be shielded with the slag from lunar and asteroidal mining activities; more elegantly, the Russian space station expert, Safronov, has suggested that an artificial van Allen Belt, or rotating electromagnetic field, around the structure, could be used to deflect charged particles. Such an idea could probably be developed further, and taken to provide on board electric power, or possibly even propulsion, of the same order as solar light sails. The recent discovery of "high" temperature superconductors opens up the possibility of economical artificial magnetospheres. The latest material can be superconducting at interplanetary temperatures, and energized with currents inversely proportional to the coil's radius. Thus a craft on its way to Mars can deploy a 1 kilometer radius fine coil of superconducting material, and energize it with a comparatively low current. Such a system not only can deflect solar charged particles, but can also extract propulsion from the solar wind, and points clearly in Safronov's direction. It is a characteristic of human inventive genius that one generation's insuperable problem becomes the next's golden opportunity.

Another source of radiation is gamma radiation, from sources in deep space-now believed to come from the galactic centre, or the interaction of rotating pairs of neutron stars. This is the latest interpretation of the investigations of the Compton Gamma Ray Observatory, launched from the Space Shuttle in 1991. In permanent settlements, protection from cosmic gamma radiation will need structural shielding-but, again, spaceship design would allow a small "storm shelter" of heavy shielding to be entered, rather like a bomb shelter of the last War, at short notice. Cosmic radiation is not considered a major issue on interplanetary flights. Solar observatories, plus suitably placed radiation monitors placed throughout the solar system, will eventually allow reasonable prediction times for radiation storms.

.The Apollo venture, and its sequel, Skylab, have thus led in a dozen short years to many advances in our understanding of human physiology, as well as many spin-offs in medical and electronic technology and materials science. These are being consolidated on the long-stay space missions of the U.S.S.R., and now the Russian Federation, as well as the 100 or so Shuttle missions to date. Most importantly, they have shown us that travel to another world and in a new medium was a feasible, practical, idea, and that humans are much more adaptable than was thought, and that myriads of tasks can be performed in space.

The Moon was shown to be geologically very exciting; and, as we shall see later, it will play an indispensable part in the future economy of Humanity in space and on Earth.

Experiments so far have shown that in materials processing, crystal manufacture, mixing of metals, and energy production, a new industrial revolution in space lies within our grasp. The era of the International Space Station will surely see the bridgeheads opened up by the pioneering work on the Space Shuttle, Mir, and Spacelabs developed into a new arena of human activity-we are only at the beginning.

Finally, the medical studies have led to a remarkable conclusion; there are no physiological limits to human activities in space which cannot be met and overcome; the main problems arise on re-adaptation to Earth, and, in any case, artificial gravity by centrifugation will prevent or reduce them to acceptable dimensions.

To those who are amazed that this totally unnatural environment is apparently congenial to humans, the answer is obvious. Most of Life has lived in a weightless ocean, the Sea. It seems implicit in the story of evolution that, armed with courage and technology, Life should flourish in the new and greater ocean of space. This proposition, so long ridiculed, has been abundantly proved in the first few years of exploration and discovery. Let us turn to the growing maturity of our movement into space, as the number of players in the game has grown, and a new industrial revolution beckons.

Chapter 6

Space and the World Community

In 1945 Arthur C.Clarke put forward the idea of a world communications network employing high orbital relays. In 1962-3 the experimental satellite Telstar relayed the first television pictures across the Atlantic. From this date the concept of the "global village" began to assume reality. The first communications satellite, Earlybird, in 1965 could relay 240 telephone calls across the Atlantic, and became the first satellite of the Intelsat series, which, has expanded in size, power, and complexity with each generation of satellite-Intelsat 8 as of 1997. By 1967 the Intelsat 3 series was in orbit over the Atlantic, Pacific, and Indian Oceans, providing the first electronic planet-wide communications. In 1971 the governing body of Intelsat represented over 100 nations, and, on its inauguration, Arthur Clarke pointed out that the signatories were signing the first articles of a United States of the Earth.

Communications has grown so rapidly that, as of 1995, the Intelsat 7 series has expanded to gigantic 4.2 ton satellites , each carrying 24,000 telephone conversations simultaneously, or 2 colour TV channels, computer and fax data. News and events are watched by hundreds of millions all over the world. The Russian Federation has its own Gorizont and Ekran systems of communications satellites, which greatly facilitate communications across that vast land. It was said in 1982, the quarter centenary of the Space Age, that, without communications satellites, the world banking system would collapse overnight.

The world wide web , together with the satellite networks, mean that, increasingly, people will belong to interest groups rather than traditional, nation States. Many large, continental states have adopted satellite technology as a means of rapidly opening up distant areas.

The Canadians have used their Anik series of domestic satellites to consolidate links between the cities of the South and remote Arctic regions. A remote medical clinic in Baffinland, equipped with a small camera, allowed remote Eskimos to be diagnosed and treated by doctors who were able to consult specialists in Toronto by satellite. Similar "telediagnosis" has been applied to transmitted electrocardiographic tracings of patients' heart rhythms. Isolated communities were able to maintain links with family and cultural facilities in the capital 3,000 kilometres to the South.

The U.S.S.R., now the Russian Federation, has carried such a system very much further and has developed mobile ground stations to link up the Far North with the facilities of metropolitan Russia. Forced prisoner labour was acknowledged as being unsuitable for the complex tasks required by a modern economy, so that skilled people have been increasingly lured by higher wages into the northern outback. Triple wages work up to a point, but for the really top class settlers, more is required. An end to cultural isolation has a lot to offer a government bent on opening up remote regions. A neurosurgeon in St.Petersburg who specializes in precise brain surgery, uses computer assisted calculations to guide his scalpel. He can assist a surgeon hundreds of miles away by using a TV camera linked by satellite. He can give numerical instructions to the local operator without having to leave his busy schedule. It is much cheaper to transmit information than human beings. Whatever the current travails of post-Communist Russia, the space-based revolution is likely to endure, and may well turn out to be the sole long-term benefit of that unhappy period. I have often felt that the mild-mannered Tsiolkovsky will be remembered long after the odious memories of Lenin and Stalin has been forgotten.

In 1974, Wernher von Braun caused to be launched a communications satellite project of profound and lasting significance. The Advanced Technology Satellite ATS6 was to carry out experiments towards direct broadcasting. This involves the use of more powerful satellites to broadcast TV to much smaller receiving stations than is normally the case. This led, in the later 1980's, to small rooftop aerials, 0.5-1 metre in diameter, removing the need for 10 meter aerials. The result was satellite TV, broadcast on a continental basis, in dozens of channels. The ATS6 satellite became the vehicle for a remarkable experiment in 1974. It was placed over India in geosynchronous orbit; 1,000 selected Indian villages were equipped with cheap black and white communal television sets, while the villagers were instructed in the building of makeshift aerials, made of chicken-wire, by local labour. Programmes on agricultural techniques, pest control, English language, family planning, basic health care, were broadcast. At the end of the year, the villagers expressed great satisfaction with the scheme. The Indians prevailed upon NASA to let them use the satellite for another four years, until 1979, when NASA resumed control. It was then used for long distance medical consultations in remote Appalachian mountain villages in the U.S.A.

It is not hard to see great significance in this Satellite Instructional TV Experiment. To bring these amenities of communications, entertainment, and education to Indian villages-550,000-without space technology would take 10-20 years, involve the training of tens of thousands of teachers, the building of schools, roads, telephone links, and so forth.

This would be very costly, and take a generation.

Space and Societies

The Indian government and NASA believe that a nationwide extension of SITE could be set up in under 1 year, and provide a wide general education for India at a cost of 1$ per person per year. No other conceivable scheme comes close. In August 1983 Ford Aerospace built for India the 1 ton Insat A to carry the telephone, meteorological, and data transmission requirements for the entire subcontinent. The failure of Insat A after a few months caused severe problems for India, with its growing reliance on satellite technology, but Insat B, launched in 1984, has filled the gap. The second generation Insat series are built by India herself, while later versions will increasingly be launched by India's own rocketry, which has already lofted the small experimental Rohini scientific satellites, as well as a polar-orbiting Earth Resources Satellite in 1994. As of 1999, India has considerable experience in the manufacture and control of indigenously built satellites, and has actually placed more into orbit than the United Kingdom. India's own Earth Resources satellite, launched by her indigenously developed polar orbit satellite launcher, made it into orbit in 1997. The administrations of Mrs.Indira Gandhi, and her son Rajiv, saw space technology as a short cut in the development of a sophisticated economy, and their successors, just like the post-Communist rulers of Russia, will continue. India even sent a "guest cosmonaut" on board the Salyut 7 space station in 1984. The United Kingdom caught up-eventually-in 1991-with Ms.Helen Sharman. Michael Foale, a NASA astronaut employee of British origin, spent several months on Mir in 1997, contributing mightily to maintaining and repairing the ageing station.

Indonesia has even greater communications problems, with several thousand islands scattered over 5,000 kilometres, so that it is not surprising to find its rulers in the forefront of those anxious to avail themselves of the Space Shuttle transportation system. As of 1984, the Palapa series of satellites with its network of small ground stations was one of the most advanced regional communications systems. There are two in service, but Palapa 3 was unfortunately lost during a mishap with 2 payload assist modules on the 10th Shuttle flight in Jan.1984. These were later recovered and relaunched.

China, too, has now established itself as a space power, with its own medium lift geosynchronous orbit satellite launcher family, domestic communications system, meteorological satellites, and also services to foreign customers. China also offers a materials processing service, in which up to 300 kilogram payloads can be flown on board a 3 ton satellite "bus", and brought back to Earth for analysis. Firms in Germany have already taken advantage of this opportunity. China is developing heavier rockets, and of a prototype crewed spaceship, Shenzhou, was test flown in November 1999. There are taikonaut training facilities, and a space station programme is now officially approved. The world's largest open university distance learning programme, with 55 hours per week of educational TV, produces 2 million graduates in China per year, from hitherto isolated villages in their tens of thousands. Worldwide, the education of millions of rural women is likely to do more than any other single factor to improve the lot of the poorer world and stabilize its runaway population growth, and would be unattainable without space technology in less than 2 generations.

In autumn 1997, Brazil launched her first four stage solid fuelled rocket. The flight failed after 2.5 minutes, but 3 other launches are expected within 2-3 years. Brazil aims for a niche in the growing small satellite launch market, and has five small satellites under development, as well as a major collaborative Earth resources satellite project with China. Brazil has also undertaken to develop equipment for the International Space Station.

These projects are have enormous long-term geopolitical consequences. The educational and communications prospects for large areas of the developing world are improved at a stroke, and a whole generation of the West's industrial revolution can be bypassed. These countries have the chance to leap straight from the age of the "bush telegraph" to the communications satellite era, by-passing land based telephones and railroads to a considerable extent. Indeed, in the coming post-industrial; "information age", hitherto poorer countries have an opportunity to get in on the ground floor, with less to restructure than some of the arthritic more "developed" economies. The chief worry for the advanced West in coming decades will be to maintain its preeminent role ; another industrial revolution will not suffice; a whole new way of technological life will be needed, especially if high-tech armaments are going to be less in evidence. A sense of unity can be fostered, as can planetary, even cosmic, consciousness. India, in particular, consists of 17 states, speaking over 100 languages. English is the nearest to a *lingua franca*, and the unity of the state is fragile at times-more especially with the conflict between Islam and militant Hinduism. The likeliest result of the comsat programme, in the long run, is an increased sense of unity among her 900 millions, together with a greater awareness of her role in the wider world. Rural isolation which has endured for millennia is breaking down rapidly. The Indian government is not unaware of this trend, nor yet of the pride which many millions of Indians feel in their burgeoning space programme. India , by 2000 is expected to be making a profit of $140 millions per year from exports of space related electronics, and data from its own space based Earth resource survey programme to, of all places, the U.S.A. If the current moves towards a more liberal market economy endure, India is capable of becoming an economic giant in the next generation, with a wealthy middle class greater than that of the whole of Europe. This is in no small measure due to her rulers' wise investment in space technology. Further, the existence of a vigorous space programme is a powerful motivator for the young to work hard at school, particularly in the sciences, technology, and engineering.

A further result of the dissemination of communications, computer networks, and education, to the remote villages of the third world could well be a reversal of the disastrous drift from the villages to the over crowded cities. Any help in checking it will repay investment handsomely.

In Africa, proposals have been under discussion between a group of African states, headed by Nigeria, and the European Space Agency, to develop AFSAT; this would be a project to set up a similar set-up to SITE for Africa-which could have enormous long-term effects on that region's stability. South Africa, newly liberated from Apartheid, would be able to lead such a programme for Africa. The benefits of the communications revolution wrought in India are sorely needed by the unhappy continent of Africa, for the problems of a conventional industrial revolution to lift standards of living are likely to prove even more intractable than in India. Many of Africa's people have paid a high price and received little reward for independence in an interdependent world.

A Growth Industry.

Estimates put the annual telecommunications market at over $100 billions per annum by the close of the century, and it is encouraging to note that Europe is very active in this area. British Aerospace developed and has test-flown the world's largest communications satellite, Olympus, in 1986. It is 17 metres in span, and provides continent-wide, or small "spot" areas, of TV broadcasting, and, with its predecessors, the European Comsat (ECS) series provides true European Television, teleconferencing, data, message, and fax transmission. In the long run, this, and the other work of the European Space Agency (E.S.A) is likely to lead to a greater cultural and industrial unity in Europe. Because the E.S.A embraces professionals from 15 nations working on large-scale transnational projects on the frontiers of science and technology, there is much greater harmony in its operations. Moves towards European Unity from this quarter will be slow and organic in nature, and thus likely to strike deeper roots than the artificial harmonization moves imposed by a hidebound bureaucracy in areas where a synthetic, uniformity is counter productive. As Dr. David Owen wisely said some years ago for the European elections, the work of the E.S.A is exactly the sort of thing for which the collaboration of the nations of Europe is intended for.

Satellite broadcasting is already making itself felt in unexpected ways. Firstly, during the August 1991 coup in Russia, the outside world, appalled by the prospect of a bunch of reactionary dead-beats seizing control of the fledgling new Russia , rallied to Boris Yeltsin's support, and, it is now widely agreed that publicized encouragement from the rest of the world emboldened Mr. Yeltsin and his supporters to overcome the coup. It is extremely difficult, if not impossible, for would-be totalitarians to cut off a country in order to cover up their crimes, or, more damagingly, their incompetence! The latest example is the Islamic Republic of Iran, which has a real problem on its hands. The youth of Teheran have developed a penchant for western TV and broadcasting, and experiment with Alien ideas and culture. The people of Iran have become a little fed up with the cultural fare offered by the Ayatollahs, and are not accepting it-the more so since ideological purity and isolation do not deliver the goods economically. There has, therefore, sprung up in Teheran a subculture of roof-top satellite aerials, which the Islamic authorities have banned. This has given the youth a considerable stimulus to their engineering ingenuity, and various disguised deployable aerials have emerged. If the West does not isolate Iran excessively over the coming years, the regime of the Ayatollahs will be softened by the cultural allure of the "Great Satan". The other possibility is that, in desperation, the Ayatollahs will use their growing military power, aided by China if she chooses the path of non-integration with the rest of the world, to make a dangerous foreign adventure.

The Age of exploration and discovery has also brought about new international co-operation and sophistication in weather forecasting, storm monitoring and prediction; more accurate and intelligible forecasts have resulted from successive generations of meteorological satellites. The World Weather Watch of 1973 employed satellites, balloons, and ships to provide a more detailed view of the world's weather than ever before. The temperatures and cloud densities at different levels can be observed over vast areas of the planet at once, and treated mathematically as a single entity. This has led to much more accurate "models" of the world's weather. Taken with the observations of solar physics from Skylab, Salyut, and Spacelab, the long-term factors influencing climate over the years will become better understood; this will be aided by supercomputers, with their ability to perform billions of calculations per second, to process the information derived from satellites. The introduction of Seasat and Topex, which can measure sea temperatures over the whole planet, the sideways looking radar which can measure wave heights to less than 10 centimetres, and the monitoring of volcanic dust output, show the interaction between Sun, solar wind, upper atmosphere, volcanoes, Ocean currents such as El Nino, and human atmospheric effects, as an integrated whole. Scientists at E.S.A speak of "Instrumenting the Grand Machine" over 20 years, with a set of several hundred satellites and probes, ranging from solar telescopes to data relay satellites, and real time space station observations, using the improved eye capabilities to detect hurricanes and typhoons at their earliest stages.

Another important development in these first years has been the Earth Resources satellites. The Land-sat series, and the Russian equivalents, can scan the Earth in many different wave lengths, and provide information about vegetation, the health of crops, the water flow of rivers, the run-off of Polar ice, and mineral deposits overlarge areas of the Earth. For instance, they can detect early potato blight 2-3 weeks before it becomes evident on the ground. This has saved hundreds of millions of dollars in wrecked crops. Warnings of impending hurricanes has saved hundreds of lives, and it has been possible to monitor the circulation of pollutant chemicals in the atmosphere, as well as the growing holes in the Antarctic and Arctic Ozone layers, and the rise in sea temperatures. Such measurements have greaty strengthened those governments pushing for environmental treaties, such as the Montreal Convention to phase out the manufacture and use of chloro-fluoro-carbon greenhouse gases. One cosmonaut on Salyut 7 remarked that a well equipped cosmonaut could achieve better results in five minutes than an aerial survey from aircraft could in 2 years. In launching the 1975-9 five year plan, President Leonid Brezhnev devoted considerable emphasis to the space programme in contributing to the advancement of the Soviet economy in all spheres. With a vast country adequate surveillance of resources, water flow, pollution and crop welfare can only be done from space platforms. Perhaps the most telling evidence of the value of the Russian space programme is the fact that Boris Yeltsin came to office as President of Russia in 1990, and the

Russian Federation at the end of 1991, committed to abolishing the entire space programme as an immoral example of Communist megalomania, with no relevance to the needs of the long-suffering people. With the level of popularity he enjoyed, he could have abolished Christmas in 1991, and survived! There was no compelling reason, other than its obvious value to the country to prevent him from carrying out his promise. The space programme is now run on a much lesser budget, and is much leaner, more market orientated, and demilitarized, but, nevertheless, they maintained the permanent crewing of the multi-modular Mir space station, which succeeded the Salyut 7 in 1987, and have methodically built up its suite of modules. Automatic re-fuelling and logistics by Progress tankers have become routine, while crews from several nationalities have worked on board the complex. What factors influenced Boris Yeltsin in continuing with a programme he reputedly abhorred? Much is of necessity speculation, but surely some of the following factors must have played a part.

Firstly, the space programme, actually contributes very greatly to the management of the national economy-to the tune of, it is estimated, at least 1 billion pounds per annum.

Secondly, the guest cosmonaut programme brings in considerable amounts of foreign currency.

Thirdly, under the Communists, and to a lesser extent, the post-communist regime, space diplomacy is a valuable tool of international relations.

Fourthly, the space programme is a vital stimulus for Russia in modernizing its economy and in bringing forward a new generation of school children interested and competent in the new sciences and technology.

Fifthly, Russia has the very real potential to create a viable and prosperous international marketplace for its space products, and, in time, for its manufactures in space factories. Mr. Yeltsin trained as an engineer , and is perhaps better able than some Western rulers to appreciate the longer view.

Sixthly, Russia has reason to fear possible nuclear and missile proliferation among desperate regimes, or even terrorists, in the dangerous years to come. One certain way to ensure Armageddon would be to allow thousands of Russian nuclear and rocket scientists to become unemployed and disgruntled-the more so since the ground-breaking disarmament treaties of the closing years of Communist rule. It is of this reason that economics and political necessity have brought America and Russia, the old space rivals, to the brink of merging their programmes in the International Space Station.

In July 1996 at a Colloquium of the International Astronomical Union on education in astronomy I had a meeting with Dr. Yekaterina Aleshkina, Director of the St.Petersburg school of astronomy, aviation, and cosmonautics and learned that youngsters in Russia still attend her courses in considerable numbers with undiminished interest.

Lastly, and, I suspect, more importantly than anyone in the West realises, is one very simple fact. In three quarters of a century of murderous and oppressive tyranny, the Russian people endured terror, famine, and declining relative standards of living-facts which now are plain to see. They have a ruined landscape, and pollution levels beyond belief. Much of their vaunted military and technological pre-eminence has gone with the Wind. Nothing Soviet appeared to work. After 75 years of tyrannical and ham-fisted incompetence, they have one achievement which they can point to as a world class success-the space programme. It is difficult to underestimate the blow to the pride of many Russians occasioned by the collapse of their imperial status , and the loss of purpose which the demise of Father Lenin must have brought to millions. Out of the wreckage the Russians can still say with pride-"We were the first in space, and no real or worthwhile advance into the new frontier can refuse to acknowledge that a Russian, Konstantin Tsiolkovsky showed the way, and a Russian engineer, Sergei Korolyev gave us the tools, and, even now that we are brought low, it is our space industry that will provide the first building blocks of the new international era". Historians will remember these achievements long after the Bolsheviks have sunk from the memory of Humankind.

Peace from Space

Many people point to the growth of military reconnaissance by the Great Powers as a negative feature of the new technology, and, indeed, the ability to read a newspaper from 500 kilometres up is a little scary. However, this facility makes it difficult for any militaristic power to hide any substantial military secrets and also to pull off any really serious military manoeuvres without detection. Many wars have resulted from a misreading of other peoples' intentions, so this is rather reassuring.

World War 1 began out of a panic resulting from escalating general mobilization, which arose from distrust. This war fever was based on insufficient understanding of each others' true intentions, and, in the initial stages was largely shadow-boxing. Similarly, Adolf Hitler achieved amazing results by bluff and counter bluff, as a result of which England and France allowed him to consider them too feeble to fight back. We overestimated Hitler's strength, and appeased him; a true and unambiguous estimate of the German forces opposing the Allies in the Rhineland in 1936 would have stiffened our resolve, and, perhaps, brought the Third Reich to an early end with minimal cost. Finally, he underestimated our determination once we resolved to stand and fight. With satellite-based espionage, it is unlikely that such a colossal mis-understandings of military resources and intentions could occur. For example, the failure of a Soyuz T launch in September 1983 was observed by American spy satellite and reported extensively in the Western Press. After the crew landed, the Tass spokesmen grudgingly admitted the failure. Without satellite reconnaissance, there could easily have been a complete cover-up. The Soyuz launch was not of particular political importance, but the lesson is clear. Verification of nuclear test ban treaties and other strategic arms agreements, depend unavoidably on the expressed use of satellite monitoring and reconnaissance. The European S.P.O.T. Earth resources satellite launched in board Ariane in 1984, made the first identification of a plume of radioactive plume of smoke from the nuclear power plant explosion at

Chernobyl in 1986; this compelled the Soviet authorities to come clean about the worst civilian nuclear disaster in history much earlier than they would have done if left to their own devices. This event, together with the surprise announcement by President Reagan in 1983 that America was to research and develop a space-based anti-missile screen are widely believed to have been proximate causes of the downfall of the Soviet Union. The new Communist leader, Mikhail Gorbachev, knowing that the Soviet economy was hopelessly arthritic with military spending and a global empire of client states, suddenly faced the appalling prospect of an arms race in space, with computer technology that the Russians, whose computer industry had been held back by the stultifying effects of Communist orthodoxy on the thinking processes of intelligent people. He could not know whether Reagan was really serious-it had only taken the Americans 8 short years to win the race to the Moon from behind. Gorbachev could not afford to take the risk of ignoring the threat, and, being fundamentally decent as well as intelligent, saw that talking and liberalization was the only feasible option. He found, as have other reformers before him, that once fear is removed from a despotic regime, the glue falls apart, and complete collapse, is the usual outcome.

When the history of the fall of the Soviet Union-surely one of the most devastating attempts to remake human nature in all of history-comes to be written, space technology will be seen to have played a not inconsiderable part.

Along with the growing impact of telecommunications, weather forecasting, resource surveys, and reconnaissance, there is a growing number of space programmes around the world. China has a capable family of launchers, both for materials processing in Low Earth Orbit, and for Comsats and weather satellites in Geosynchronous orbit. Earth resource satellites are underway, as is international marketing of launch services. We can expect considerable increase in the range and quantity of Chinese space activity, and, depending on the political situation after the death of Deng Xiao-ping, crewed spaceflights on indigenous vehicles are likely over the next decade. Indeed, as of September 2000, a programme for crewed flights leading to a small spece station by 2010 has been officially approved and announced.

India is moving towards home-grown Geostationary and polar orbital launchers and payloads by the year 2,000. Intentions as to Space Shuttle development of her own is much more speculative, but is not ruled out by Indian space engineers. An unmanned lunar mission is also being discussed.

Israel is now a space power in her own right, with 3 Ofeq satellites launched as of 1995. These are research satellites, weighing up to 220 kilograms, and are leading to an Israeli astrophysical satellite, and a reconnaissance/communications satellite within 3 years.

The ability to detect an impending sneak attack from any of the rejectionist neighbouring states-such as Iran, or an Islamic coalition led by her- is of great importance to Israel, and, indeed, the rest of us, since any realistic prospect of defeat could trigger the "Samson Option"-the nuclear card. The result of this particular miscalculation could be devastating for civilization as a whole. An early warning from space could literally be vital to all of us.

Japan has become a major player in the space league, with a 25 year history of scientific satellites, including probes to Halley's Comet, and the first, apart from America and Russia, to send a probe to the Moon. Japanese astronauts have flown, and are being trained for Shuttle and Space Station operations. Japan is building a Laboratory module for Space Station , along with the European Space Agency. In 1992, Japan introduced her new heavy H2 rocket, which can place 2 tons into geosynchronous orbit, and 10 into Low Earth Orbit. An improved version, the H2A, is being developed to carry 3 ton satellites to GEO, at reduced cost. She is also developing a miniature Space Shuttle, called HOPE, to be launched by the improved H2A, after 2004 AD. Japan is conducting automatic docking and rendezvous experiments, and envisages a single stage to orbit re-usable rocket developed from H2A and HOPE technologies to be in service by 2010, operating at $1000 per kilo of payload to Low Earth orbit with a Skylon type airbreathing hybrid vehicle by 2025. Japan also is developing unmanned lunar, Mars, and asteroid missions over the 1999-2006 period. Japan has taken the challenges of space and its opportunities seriously; indeed the construction firm Shimizu, employs 30 people full-time to design orbiting space stations, hotels for space tourists, and Lunar bases for the next 30 years, in the full and clear expectation that, they, Shimizu, will be building them.

Up to 1985, in their second generation space stations Salyut 6 and 7, the former Soviet Union had flown guest cosmonauts from up to 10 different nations for missions, typically, of 8 days or so. These comprised the old Warsaw Pact nations, plus France, India, Cuba, Vietnam, and Afghanistan. Their third generation station, Mir, carried this trend further, with some longer stay guest cosmonauts, such as Jean-Loup Chretien from France, who stayed 1 month, as well as the American Doctor Norman Thagard, who has completed a 110 day flight in summer 1995. Astronauts from the E.S.A. spent 1 month in 1994, and 4.5 months in 1995. Many experiments have been carried out by the guest cosmonauts; for example, Arnaldo Mendez of Cuba tested the Cuban boots, Vladimir Remek tested the ""Kristall" electric furnace for the smelting of exotic alloys, East Germany's Siegfried Jaehn tested a new Earth Resources camera, while the Frenchman Jean-Loup Chretien performed blood lymphocyte measurements before, during, and after spaceflight. He also revolutionized food in space by insisting on French cuisine on board Salyut 6. It is doubtful whether the chefs at Tyuratam ever got over the shock-more was to come from the Indian guest in 1984! Continuing international flights on Mir have provided much valued foreign exchange for the troubled post-Communist Russian Federation, and also reinforced Russia's claim to great power status in difficult times.

America has launched women, coloured Americans, and many foreign nationals, notably Canadians, Japanese, Dutch, Germans, Australian, and many others, as of 1995, over 20 different nations have sent their nationals into space, and over 360 human beings have flown. This is the true internationalization of space, and, it is to be hoped, will begin to have beneficial effects on Earth by bringing the brightest and best of all nations to work together in common, challenging tasks. The Space Shuttle's capacity to retrieve and repair satellites, and to operate reusable and flexible space laboratories, has already been put to good effect for several nations, while the scientific and cultural legacy of the Great Observatories is growing in importance with each passing year.

Europe comes of Age

It is time to look at the emergence of the largest, and most commercially successful space programme outside the traditional "Big Two", namely that of Europe under the direction of the European Space Agency (E.S.A). The E.S.A came of age during the 1980's, after its formation from the amalgamation of the European Launcher Development Organization and the European Space Research Organization (E.L.D.O & E.S.R.O) in 1973. The omens at first were not good. E.L.D.O. had signally failed to develop a European launcher based on the Blue Streak missile as a first stage. This last had worked well, but the three stages in combination had never performed satisfactorily. Its initial thrust was 136,000kg. and was considered insufficient for the anticipated heavy communications satellites of the 1980's and 1990's. There were 11 signatory nations-now 15- being all the members of the European Union except Greece, but including Switzerland. Each country agrees to fund the E.S.A budget according to a percentage, but to allocate it to specific projects according to national priorities. The United Kingdom's main contributions have been in the area of Earth Resources, weather, and communications satellites, as well as the highly successful scientific satellites, notably the Giotto probe to Halley's Comet in 1986, which was redirected to Comet Grigg-Skjellerup in 1992. The U.K's enthusiasm for space activity has notably declined in recent years-after contributing 2.5 % to the development of the Ariane 1-4 rockets, the government in 1987 refused to participate in the next generation Ariane 5 launcher, or the Columbus space station module, despite the fact that Ariane has now captured 50 % of the comsat launcher market, and has repaid its development costs many times over.

There are three main components to E.S.A's programme. First, it was agreed to develop an all-European satellite launcher. This was to be a heavy three-stage launcher specifically tailored for the geosynchronous orbit, where the most business was anticipated. (At least 60% of the development costs and work was done by France through its own C.N.E.S space industry, reflecting a long-standing French ambition to have independent access to Earth orbit. The Blue Streak experience was built on, and the launcher, named Ariane, was considerably more powerful, developing, in its early basic versions, some 250,000 kg of thrust. This version, Ariane 1, had four main Viking engines as first stage, and could put 1.9 tons into geosynchronous orbit. Major launches included the Maritime comsats Marecs A & B, the European Communications Satellite E.C.S, Intelsat V, numbers 7 and 8, and, most famously, Europe's first "interplanetary" satellite, Giotto. By 1984, America was beginning to use the Ariane launchers for some of their commercial launches-and by the end of the 1980's Ariane was carrying 50% of the world's geosynchronous traffic, with its order books filled 2 years in advance. By uprating the third stage, and adding two solid fuelled strap-on boosters, Ariane 3 raised its lifting power some 50% to 2.7 tons, while, by 1988 the latest version, Ariane 4, flying in up to 6 combinations of pairs of solid and liquid strap-on boosters, raised the maximum payload to Geosynchronous orbit to 4.4 tons. By the late 1980's, therefore, Europe had acquired a powerful and flexible independent launcher, and was ready to contemplate the next steps. In 1984, E.S.A handed over the launch facilities and the purchase, launch, manufacture, and marketing of Ariane to a private Europe-wide consortium, Arianespace, which consists of 36 Aerospace companies throughout Europe. Europe, in effect, has created the first commercial launcher business, and, during a period when most of Europe's industry has been hit by recession, it has put up an astonishingly successful performance. There are no grumbles from the European tax payers about money being wasted in space. It is worth noting that the Ariane 4, if reconfigured, could place 8-10 tons or so in low Earth orbit, and is slightly more powerful than Russia's A2 Soyuz launcher.

Another major European project is the Spacelab. This was the result of an agreement between N.A.S.A and E.S.A. in 1973. N.A.S.A agreed to buy two scientific laboratories for space research at a cost of $500 millions. This was to be a highly adaptable 4 man "shirtsleeve" environment with varying combinations of pressurized cylinders for human laboratory work, and U shaped external pallets for experimental work requiring exposure to hard vacuum, such as physics and astronomy. It could consist of two living modules and one pallet, or 5 pallets on their own without a crew, or one living module with 3 pallets; these combinations could be altered according to mission requirements. In a full biological/Life sciences mission, the result was a cylinder 4 metres in diameter and 7 metres long, in which four people could work. The main contractor is Germany, which pays 60% of the development costs. Under the N.A.S.A /E.S.A agreement, E.S.A provided 60% of the experiments on board Spacelab, and European astronauts had equal access to the lab. It has flown at least every 2 years since 1983, and been very successful as a scientific laboratory, with research programmes being modified while flights are in progress to take advantage of results as they come in. The main limitation of mission durations has been the lack of electrical power, and the limited life support system of the Space Shuttle, which was designed purely to provide access to space.

The Spacelab was equipped with advanced data handling and processing equipment, which could deliver information to the Earth via a highly sophisticated communications satellite-T.D.R.S.S-at a rate of 48 Megabits per second. This is approximately equal to 15,000 pages of print, and allowed scientists to have access to experimental results in "real time".

Instrument packages could be flown on the Shuttle, returned to Earth, and rejigged for follow-up missions two years later with improved instruments. This allowed for long-term, on-going, scientific and industrial research programmes, many of which will bear fruit when permanent access to space stations becomes routine. As with the history of science and industry on Earth, a whole range of applications and processes will become available. The Spacelab mission carried 73 experiments of a wide-ranging nature; these comprised five main areas-life and medical sciences, atmospheric research, mapping and geography, solar and astrophysics, and metallurgy. In medicine, studies of the space-sickness syndrome were made by using elastic cords to produce simulated pulling and pushing forces, electrical stimuli, and a special rotating chair, developed to produce precisely tailored rotational forces.

Studies of cardiac rhythm and force of action were made with two new instruments; firstly, a three dimensional ballistic echo-cardiograph measures the dynamics of the heart by detecting the slight reaction of the astronaut's whole body to the heart's pumping. The body moves in an opposite direction to the cardiac pumping by a minute amount according to Newton's Third Law; this is the same principle as that governing the movement of rockets in space, as described in Chapter 4. Secondly, a portable personal electrocardiograph, opthalmograph, and encephalograph, developed in Britain was used in Spacelab. Also, an ingenious British experiment to test human perception of mass in space was made; here,12 identical balls were used, of identical appearance, but different mass. Studies of sunflower seed germination, microbial growth, exposure to cosmic rays, and blood lymphocyte activity were made.

In the other sciences, measurements were made of the output of solar energy, the formation of aurorae with charged solar particles, and the levels of atmospheric contaminants. Detailed mapping of large areas of the Earth with new special cameras was carried out-many parts of the Earth had not hitherto been mapped accurately with modern instruments. The study of solar light absorption, and the scattering and dispersal of light by chemical contaminants will have far-reaching implications for our understanding of climate, and its interaction with factors such as volcanic dust output as in the Mount St.Helen's and El Chicon eruptions, and the effects of pollution.

With the operation of the Ariane rockets, wide-ranging satellites in the scientific research, as well as the commercial arena, and the beginnings of human space flight experience, the E.S.A had come of age by the late 1980's, and was ready to consider more ambitious steps. In 1987, the European Council of Ministers relevant to space matters met to consider three main proposals;

First, the development of a new, larger, Ariane 5 heavy booster, to be capable of putting 6-7 tons into geostationary orbit-or 3 large satellites. This vehicle was to have a cryogenic hydrogen/oxygen first stage, flanked by two 250,000 kg thrust solid rocket boosters. In a three-stage version, Ariane 5 is to place 6-7 tons into Geosynchronous orbit, at a cost per kilogram a mere 40% of the original Ariane 1 of 1979-84. In a two stage version, Ariane 5 is to be capable of 18-20 tons to low Earth orbit. This will provide the ability to place space station components into orbit, of the same size as the Russian Mir base block, or principal add-on modules.

Second, E.S.A proposed a space station module, building on the experience of Spacelab, to be called Columbus, which would initially fly attached to the planned International Space Station, as designed in 1984 for launch in 1993-4. It was envisaged that around the turn of the century, Columbus would be detached from the American complex and become an independent, free-flying, European Space Station, to be serviced by a European mini-Shuttle craft, called Hermes-which was item 3 on the menu.

Thirdly, France proposed a European Shuttle, Hermes, which was to carry 3-4 astronauts and 3 tons of baggage twice a year to support the Columbus programme. Hermes was to weigh 20 tons, and be carried into low Earth orbit by Ariane 5.

Two factors led to the French interest in Hermes-these were, firstly, that many Europeans felt that the equal access prescribed by the N.A.S.A /E.S.A deal was not being achieved in practice, and that Europe should have its own access, and secondly, that the Challenger disaster in 1986, which killed 7 astronauts and wrote off one of four Shuttle craft, and the consequent retrenchment of the American Shuttle programme, with the repeated down-scaling of President Reagan's original Space Station proposals during the 1980'and early 1990's made America an unreliable partner for long-term co-operation on a space station.

However, it became evident that it was simply not possible to accomplish the design goals within the weight constraints of the Ariane 5 rocket, and the projected costs kept on rising, so that, as of 1993, the Hermes project became effectively dead. This leaves Europe with no choice, for the next 10 years at least, but to be part of the International Space Station project, now renamed from Freedom to ISS. Accordingly, therefore, the Columbus programme is still underway, in a scaled down form, in that E.S.A is to provide an attached crewed module to be berthed with ISS in about 2003. This is designed to be launched by either Ariane 5 or the U.S.Shuttle, according to scheduling considerations. Germany and Italy are carrying out most of the work.

Ariane 5 was approved without difficulty, except for the non-participation of the United Kingdom, thanks to Mr. Kenneth Clarke, the then responsible minister. Before its first flight, it has already generated 12.5 billion francs in business, not a centime of which will accrue to non-participating Britain. After a disastrous maiden flight in June 1996, due to faulty software, Ariane 5 made a successful test flights in 1997 and 1998, and is now qualified for commercial operations by Arianespace. Plans for evolution to a 9 ton capacity to geostationary transfer orbit by 2003 are being advanced, while a new cryogenic transfer stage should increase its payload to 11.3 tons by 2007. Ariane 5 is expected to be Europe's workhorse until at least 2015. ESA envisages evolution towards partial and then full re-usability, with an ultimate single stage to orbit vehicle, to be based on the Ariane 5 system, by about 2020-5.

With regard to the possible European crewed space ship, consideration is being given to a small Apollo/ Soyuz type 3-4 man vehicle, which could be developed at a cost of c. $1 billion, and launched by an Ariane 5, suitably upgraded to crewed safety standards. This could be available within 4 years, if required. For the further future, E.S.A began a programme called Future European Space Transportation Investigation Programme, which is, over the next 2-3 years, to consider single or two stage fully re-usable vehicles. In 1984, Alan Bond of British Aerospace produced a plan for a re-usable space-plane, called HOrizontal Take Off & Landing (H.O.T.O.L), which would take off on conventional jet engines, and, at 13,000 metres or so, after passing through 95% of the atmosphere, switch them over to internal oxygen supplies, thus converting into a high performance rocket. This proposal was initially funded for two years by the United Kingdom, but not with sufficient conviction to convince our E.S.A partners to support it. Wind tunnel model tests, and materials studies, showed that the concept was likely to work rather better than Bond predicted, and that a 230 tons space plane could take 8-9 tons to Low Earth Orbit. (Improved versions, S.A.T.A.N and Skylon followed). Indeed, Skylon has now been designed to carry 12 tons into Low Earth Orbit, twice a week, at less than $14 millions per flight, with a 275 ton fuelled up launch mass. Alan Bond and Richard Varvill have formed a company, Reaction Engines Ltd, to put together as consortium to develop Skylon further, but have been characteristically neglected by the British Government. Presumably, they are hoping to buy it cheaply from the Japanese in about 2015. However, the new Labour government was said to be slightly more receptive to the case for Skylon, especially as Bond Working Group has put together a team of international financial and industrial backers who require only $28 millions over 18 months from a UK government to persuade them to develop it privately. They have identified a market for 35 vehicles, at $550 millions each. Germany has on its drawing boards a two-stage space plane, Saenger, whose first stage could well be developed as a Concorde successor. Such vehicles are likely to be of interest to the FESTIP Programme over the next 10 years.

Meanwhile, E.S.A is preparing for the space station era by building up its experience of crewed spaceflight by sending astronauts on board

Mir and the Shuttle, and have already gained considerable expertise. Having seen the European Space Agency mature from formation in 1973 to maturity as a first class player in nearly all the aspects of space development, a brief diversion into politics is perhaps appropriate here. The high ideals of the founding fathers of the European Union have been bogged down by the conflicting national interests of the member states, while the Commission in Brussels has earned great unpopularity by trying to impose harmonization in matters where this is unimportant and unwelcome. Many Europeans apart from the British resent the vast sums spent in support of small inefficient farms which produce mountains of food in excess of market requirements-or, even more crazily, not to produce food at all. Over two thirds of the Community budget is consumed by this wasteful exercise, which is, moreover, riddled with fraud. Meanwhile the real pan-European issues of defence, foreign policy, and long-term industrial and post-industrial development languish, with the relationships necessary for addressing them perpetually poisoned by the albatross of agriculture.

In contrast to the Common Agricultural Policy, the European Space Agency is a model of harmonious cooperation. Officials and scientists working in the E.S.A have differences of opinion, but these are rarely muddied by primitive nationalism, and the comradeship of shared and exciting enterprises prevails. The results, contrary to earlier misgivings, have been impressive, and much more can be done. If the E.S.A could dispose of even one quarter of the budget currently dissipated on agricultural support, politically correct interference, and pursuing unnecessary regulations, the frontiers of space could be pushed much further back, the industrial and economic prospects for Europe much improved. E.S.A, in contrast to the United States and Russia, has a minimal military role in space, and therefore there is no room for protest by pacifist or environmental lobbies on grounds of threats to world peace, real or imagined. Cooperation between member states is essential for results to be achieved, and would restore the idealism and popularity of the European ideal. An image of Europe as a dynamic, forward-looking, harmonious and constructive enterprise, working along with America, Russia, Japan, India, and other countries in expanding the future for humanity would be far more attractive to young Europeans than the conservative, wrangling, petty bureaucratic reality of today's newspapers. President Mitterand recognized this when he called for concerted European action in the industrialization of space years ago. In 1996, for the first time, non-governmental agencies accounted for more than half the global turnover of space business, which totalled $77 billions per year in 1997.

Future Perspectives

What are the prospects for Europe in space over the next 10-20 years? In the near term, Europe will gradually replace the Ariane 4 rocket with the larger Ariane 5. There is an active programme of scientific satellites, ranging from the powerful Envisats, to follow on from the SPOT series of Earth Resource satellites, in the monitoring of the planetary environment at unprecedented levels of detail. Britain is playing a lead role in this programme. Then there are the SOHO solar, Infrared and Xray Observatories, launched in 1995-9, which will maintain a front line European contribution to astrophysics, as well as the lander component of the Cassini probe, being jointly developed with America, which set off for Saturn in 1997. The Columbus programme will be Europe' contribution to the International Space Station, while continuing joint crewed missions will build up Europe's experience in human spaceflight. The Automatic Transfer Vehicle(ATV) programme had this year won the largest ever contract for the European Space Industry-to develop and launch, with Ariane 5 rockets, a supply/manoeuvring vehicle to the International Space Station every 18 months from 2003-2014. The contract is worth some 9 billion euros, and represents a major investment and challenge for Europe. In 2003/4 ESA plans a Mars lander/rover mission, called Mars Express. Scientists from the UK's Open University, and Leicester University, are designing a drill to take subsurface samples which should throw further light on the existence of underground water, and perhaps, life on that planet. The Rosetta mission of 2004 is aiming to sample the nucleus of Comet Wirtannen.

In June 1994, in Switzerland, E.S.A produced a four phase 20 year programme of lunar exploration, beginning with an orbiting survey craft and lander in the years 1997-2002, leading via rovers and sample return missions, untended lunar bases to a permanently manned lunar base within 20 years. Each 5 year section would lead on to the next phase without a major commitment to design or construction of the final phase, so that politicians need not fear a massive open-ended budgetary commitment at the start. Other nations would be invited to contribute after phase one. The first 2 phases, at least, could be supported by the Ariane 5 launcher and the Columbus programme, while experience gained on Columbus and its support would facilitate the design of lunar habitats and transfer vehicles. If Columbus goes ahead, and the preceding lunar programme is adopted, the FESTIP is very likely to lead to an advanced space transportation system on the way, since these programmes imply a long-term European commitment to front-line space activity for decades to come. E.S.A is likely to function largely as a research and development agency, while the business of operating and building hardware will be contracted to Arianespace. The contracting governments and industries would buy services from Arianespace, or any other space operating company.

In America, there are already several fledgling private space operators, some of whom may well be the space age equivalent of Panam and TWA in the next century. Indeed, Dan Goldin remarked at the IAF Congress in Turin that 1996/7 was the first year in which commercial space turnover exceeded the government's space budget- surely a marker for the future. What of the International Space Station ? It began life in 1984, with President Reagan's invitation to the nations of the West to come together to design an international space station by 1992, which would act as laboratory, factory, fuel depot, and communications/logistic hub for the exploration and development of the solar system. Matters got off to a good start until, in January 1986, the Space Shuttle Challenger exploded in a clear blue sky with the loss of the ship and all hands. With enormous courage and frankness, they set about finding and publicizing the causes of the debacle, redesigning the system to correct them, and replacing the lost Challenger with a considerably improved product. It is clear, in retrospect, that America, government and people, passed a kind of "Rite of Passage" with the Challenger Disaster, and have now accepted their destiny in space at a very deep level. What has been bought by the loss of 7 brave individuals will not be lightly be thrown away. This being so, however, the International Space Station has had a very chequered career. It has been through at least 7 redesigns but has now found an unexpected saviour- the Russian Federation. Just as, in 1972, N.A.S.A only pushed the Space Shuttle through Congress, in a truncated form, because of the 15% international contribution represented by the European Space Agency, so the Space Station achieved its biggest endorsement in 1994, after Russia and America agreed to merge their space station programmes in the latest redesign, International Space Station. The first step took place in the summer of 1995, with the docking and crew exchange between Mir and Shuttle Atlantis-the first of 9 such link-ups before work started on the space station in 1998/9. After many delays and vicissitudes the stage is now set as of September 2000, for the realization of this massive project. All being well, by the end of 2001 the International Space Station should be the largest and most complex human enterprise in Space, and its growth will be less than half complete. It is already visible to casual observers of the night sky as a rival to the brightest stars , at the correct times and locations. These are available from the N.A.S.A space station web site

The human move into space began as a deadly rivalry-the "Space Race"- between two antagonistic powers, while the continuation of the story sees the two erstwhile rivals collaborating, and the next step heavily dependent, not on rivalry, but on cooperation. What a good omen for the Millennium! With the construction of the International Space Station the first Space Age comes to an end, and the second begins-for, unlike Apollo, which to most Americans finished when they won the race, ISS is a beginning.

To conclude, the new system of communications and education, the improved global view of our weather and climate, the backing of reconnaissance satellites for international treaties have laid the foundation for a new planetary community and consciousness. The exciting prospects of space stations, and lunar bases, permanently crewed by talented men and women of all nations, engaged in furthering the advancement of human horizons and capacities, promise further cooperation among the nations of Earth. ISS, if it survives the crises in Russia, apart from contributing massively towards the human move into space, will bring together at least 17 advanced nations into a common industrial and technical enterprise, spanning at least 30 years. On this level alone, it is unprecedented in all of history, and if the building of a World Community in science, industry and technology were its sole achievement, there would be no cause to regret the expense.

Before, however, considering the tasks to be performed on the Space Station, we shall see how, in the past 30-40 years of the Space Age, the period of exploration and discovery has shown us a whole new Universe. In such a short period we have not just ventured tentatively outside the Earth's atmosphere, nor even merely undertaken a few short journeys to our immediate neighbour, but have sent our eyes and ears to most of the other worlds of our solar system-so that, of the planets, only farthest Pluto remains unvisited. The results have been a period of exploration and discovery unequalled since the time of Copernicus and Columbus, and serves to remind us that there is far more to this story than the cooperative economic efforts of Humankind, important though these are. For the first time the children of Earth have reached beyond the cradle of their birth, and there is no return to past isolation.

Chapter 7

The New Copernicus

Many of my generation can well remember the encyclopaedias of our childhood in which the planets were described tentatively in short paragraphs with some very speculative artists' impressions. The diameter, orbital period, and distance from the Sun were listed accurately, with a thumb nail sketch of local conditions heavily hedged with suppositions. These were based on telescopic observations undertaken at the limits of visibility, with atmospheric opacity granting the persistent astronomer mere glimpses in a life-time's work. Theories of the formation of the solar system were vague in the extreme and, for most astronomers, in one of the most exciting features of the Age of Exploration, the planets have become real places with "geography", atmospheres, climates and history; they have acquired place names and separate identities.

In the 1950's even the Moon, only 400,000 kilometres away was poorly understood. It was not known whether the craters were formed predominantly by volcanic action, or by meteorite bombardment. Nor yet was it known whether the dark areas or Maria were covered in dust, or were smooth and solid. This matter was of some importance to those considering a human landing! Until the Rangers crashed into the lunar surface in 1961-4, this question was unresolved. The age and history of the Moon were unknown. Some felt that it was captured by the Earth after formation elsewhere, while others believed that Earth and Moon had condensed from the same dust cloud. Others even fancifully suggested that the Moon was wrenched from the Pacific Ocean during the early years of the Solar

System. The far side of the Moon, of course, was unknown and invisible. From the Apollo rocks, it is now clear that a Mars sized object actually collided with the Earth during its early molten stage, and that a large chunk of the molten upper regions of the Earth's crust and mantle, together with that of the colliding object, became detached from a dumb-bell shaped globule, and formed the Moon. This is clear from the fact that the surface of the Moon resembles the Earth in many of its lighter elements; but the absence of an iron core shows that there was no admixture with the Earth's core. The collision was not sufficiently violent to actually break up the Earth, which was still molten or semi-molten at the time.

The Luna 3 space probe launched by the U.S.S.R in 1959 returned the first pictures ever taken of the lunar far side, and showed it to be more mountainous and crater-ridden even than the side we can see. There is a heavy Russian influence in the naming of the far side-Tsiolkovsky has at last been honoured by giving his name to a particularly large and spectacular crater.

The Ranger and Surveyor spacecraft from 1961-7 showed a lunar surface even more pock-marked by craters than we had supposed, with a firm surface covered with only a thin layer of dust. This paved the way for Apollo. The Americans undertook 6 crewed Apollo missions to the Moon from 1969-72, as well as the dramatic Apollo 13, and brought back over 350 kilograms of Moon rock and dust from carefully selected sites. The U.S.S.R brought back about half a kilogram in two automatic sample return missions. The Soviet achievement with automatic probes has been very commendable, but it is abundantly clear that, in terms of quantity, variety, and flexibility of sample collection, and scientific activity, there is no possible question of replacing humans in space by robots for a very long time indeed, and, even if there were, robots would only serve to whet the appetites of would-be travellers. Only a small part of this material has been analysed in depth, and samples have come from the plains, mountains, rilles and crevasses-all from regions near the lunar equator. None has come from the polar regions, where some astronomers believe that, in the perpetually unlit deeper craters, deposits of water ice left by cometary impacts may exist. The 1994 spacecraft, Clementine, surveyed the Polar region, and sent back over 1 million images. Lunar Prospector, launched in Jan. 1998, at a cost of $63 millions inclusive of launch, found stronger evidence of lunar water ice in March 1998, and, by September 1998, possible quantities of up to 10 billions tons were being mentioned by Prospector scientists.

. The discovery of similar possible water ice deposits in the polar regions on-of all places-the planet Mercury-in 1994 would make lunar ice much more likely. Confirmation would greatly enhance the prospects for a lunar base.

The Moon has been dated at 4,550 million years, or nearly contemporary with the Earth, and its surface material is fairly similar to terrestrial basalt. There is more titanium and aluminium than on Earth, and similar amounts of silicon; these are to be found as oxides, so that fully 40 % by weight of the lunar surface is oxygen. There is no atmosphere, no magnetic field, and much less iron than on the Earth. Crater systems are more numerous and elaborate than expected. There are examples of younger craters within the floors of older ones, and it is accepted that the formation of Earth and Moon appeared at a very similar time, close to the origin of the solar system.

The craters on the Moon and elsewhere in the solar system point to a period of intense bombardment by "planetesimals" or large asteroid fragments left over from the formation of the planets, culminating at about 3,9 billion years ago, when the *Mare Imbrium* was created, with much less frequent bombardment in subsequent ages. After the dust had settled, the interior cooled, and became to all intents and purposes geologically dead. The lower density of the Moon reflects a lesser interior concentration of nickel, iron and radioactive elements than on the Earth. This tends to confirm the current opinion that the Moon and Earth were not formed from the same condensation, but that the Moon was condensed from terrestrial surface material before complete differentiation had taken place on

Earth, by collision with an outside object. Bombardment of the Moon, and Earth by planetesimals has not, of course, ceased entirely, as is clear from the fate of the dinosaurs. The crater Tycho, with its magnificent ray system, is dated at about 170 million years, while the small crater Bruno, on the far side near its junction with the visible limb, is very recent indeed.

This brief and unavoidably very rough sketch of the Moon already indicates that the Apollo missions were more than just an expensive version of climbing mount Everest. The Moon figures prominently in plans to use the resources of space to solve the looming crisis of resources and energy on this planet, and will become a key part of a remarkable new extraterrestrial civilization during the 21st century.

On to the Planets

The face of the planet Mercury was completely unknown before 1970's. Owing to its closeness to the Sun, it is only observable by telescope for a few days of the year when it subtends a right angle to the East of the Sun with the Earth at sunset, and to the West at sunrise. Its size and orbit were known, but even the length of its day was a mystery. Studies by radar in 1965 had shown it to be 59 days, or 2/3 of its period of 88 days, but only the Mariner 10 space craft was able to tell us anything of the planet's nature. Mariner 10 was launched towards Venus in 1973, and in an amazing procedure known as "gravity assist" used Venus' gravitational pull to deflect it inwards towards the Sun to an encounter with the planet Mercury. Mariner flew past Mercury, then cut across Mercury's orbit to encounter it again in 6 months; after a parabolic orbit around the Sun, it returned to Mercury's orbit on the return limb of its parabola. Mercury had in the meantime described two orbits of the Sun in its 6 months. A third encounter took place 6 months later, 45% of its surface was photographed, and some description of the planet is now possible.

The planet has an exceedingly long day and night, with a temperature variation of 350C to -200C. There is a minimal atmosphere, and the surface is not dissimilar to the Moon's with the same story of early heavy bombardment by planetesimals and meteorites. There is evidence of a weakly magnetic iron core, and of perpetually shadowed polar regions containing polar ice caps. This totally unexpected announcement was made in 1994, after careful study of radar images, made by bouncing radar transmissions from radio-telescopes. It seems unlikely to be a popular resort except for solar physicists or would-be industrial manufacturers of anti-matter for interplanetary trading ships, or early interstellar missions, of the mid 21st century. Before readers laugh, there have been significant developments in the subject of anti-matter and its handling in the past few years which make the above more than a gleam in the eye. The European Space Agency, and N.A.S.A have both listed Mercury as a desirable target for a possible orbiter mission in the next decade, to complete the mapping begun by Mariner 10.

Venus, the nearest planet, was always considered as Earth's sister planet, being 11,500 kilometres in diameter and, apparently, of similar density. In the early 1950's some had hoped that it would be covered with balmy tropical seas and even, perhaps, inhabited. The truth was shrouded in mystery because of the exceedingly reflective clouds which gave no hint whatever of the underlying surface. The spectroscope-a device for measuring the absorption of light of different wave-lengths, and hence of giving information about the chemical composition and temperatures of distant atmospheres-gave baleful hints that Venus was after all a wicked sister. Firstly, the infra-red absorption pattern showed that the planet had a very high surface temperature and that the atmosphere was rich in carbon dioxide. In 1967, the Russians landed the first successful vehicle on to the surface of another planet. The probe was soon put *hors de combat* by the extreme conditions on the surface, but not before it had spilt the beans. The surface temperature was 450-500 C, and the pressure, at 90 atmospheres, roughly equalled that to be found at depths of half a mile in the oceans of Earth. Mid-day on Venus is gloomy, and like a cloudy twilight in a rain forest. The high carbon dioxide content of the atmosphere was confirmed. Subsequent probes-Venera 7 to 10-returned television pictures for about 60 minutes, and Venera 11 and 12 took core samples by drilling. The last members of the series, 15, 16 and the two Russian Halley-Venus probes, included orbiters which mapped the surface by radar. That task has been very comprehensively done at high resolution by the N.A.S.A probe, Magellan, which, over 3 years from 1989, mapped Venus with synthetic aperture radar down to a few hundred metres resolution, over the course of four complete polar orbital surveys. 0.99% of the surface has been mapped in relief, and the principal geological formations have been clearly shown in false-colour images. Owners of personal computers can now buy C.D ROMs of the planets Venus and Mars, as well as of the Voyager missions to the outer planets, and explore the terrain of the hitherto totally shrouded Venus in their own homes! Venus has lofty plateaus, volcanos, tesserae, and domes of basalt, but no plate tectonics as we know them on Earth. The atmosphere, in addition to carbon dioxide, contains appreciable quantities of sulphuric acid. Lightning, and volcanic eruptions, have been observed by *Magellan*, and, of course, no trace whatsoever of water.

The vital question, of course, is; why is Venus such an inhospitable place after such an apparent resemblance to the Earth? The two planets are nearly identical in size and density, and have a similar composition to Earth's volcanic rocks; they are also only 40 million kilometres apart at their closest. They should be similar. The difference is generally explained as being due to a "runaway greenhouse effect"; the story goes as follows;- in the beginning, Earth, Venus, and Mars began with a similar amount of carbon dioxide, some of it free in the atmosphere and some of it bound to the rocks as carbonates. Each planet was also endowed with a water ocean, possibly derived from cometary bombardment, as well as volcanic outgassing. As Earth cooled, the rocks absorbed much of the carbon dioxide to form carbonate rocks, steam condensed into water, rains and oceans, which dissolved more carbon dioxide, leading to the formation of carbonate rocks and minerals. To cap it all, at just the right temperature, Life evolved, and the plant kingdom removed more carbon dioxide, and converted it into oxygen and carbohydrates, which became buried in dead vegetation. This meant that Earth's initially high atmospheric carbon dioxide, outgassed from volcanoes and during the bombardment phase, has been almost totally removed from the atmosphere, and bound up either in vegetation and rocks, or dissolved in the oceans.

On Venus, however, the case was different; after the first few hundred million years, the course of evolution took another path. Being nearer to the Sun, the Venusian atmosphere was a hotter and cooled less rapidly, so that less carbon dioxide combined with the surface rocks, and the atmospheric concentration of carbon dioxide remained high. This allowed an excess of heat in the form of solar infra-red radiation to be trapped by the atmosphere, and so not radiated back into space, as is the case during terrestrial nights. Thus, initially, the Venusian atmosphere was hotter because of its proximity to the Sun. At two thirds of the Earth's distance from the Sun, Venus receives nearly twice as much energy from the Sun. This sufficed to prevent the carbon dioxide from forming rocky carbonates, and the atmospheric water vapour remained in the guise of hot steam high in the atmosphere unable to form liquid carbonic acid. This still further reduced the chances of removing carbon dioxide from the atmosphere of Venus, and set up a vicious circle, as raised temperatures led to the release of what little carbon dioxide had been bound as carbonate bearing rocks. The early ocean, too, had a short life, since the hot water vapour was dissociated into its constituent hydrogen and oxygen molecules by solar ultraviolet radiation, and the hydrogen, being hot and light, rapidly escaped from the planet into space, leaving the oxygen to react fiercely with the unweathered rocks. Venus thus soon lacked two main stabilizing factors for its temperature. If Life ever appeared maybe 4 billion years ago, when Venus still had an ocean, it was rapidly snuffed out as the ocean vapourized, and the temperature rose to unbearable levels. Thus the best that can be said of Life on Venus is that it was still-born.

Lessons for Earth

There are two lessons to be learned here, of considerable importance for our future. In the first place, the environmental crisis on Earth assumes a dreadful importance, with the implied warning that, if we destroy the great forests of the Amazon and the Northern regions of Russia and Canada, the level of atmospheric carbon dioxide will rise, and the level of oxygen will fall, thus causing a rise in atmospheric temperature. All combustion of fossil fuels releases extra carbon dioxide into the atmosphere as well as consuming oxygen. Our industrial enterprises are considerable, and, if the above two factors are taken together, we will begin to affect the climate-some believe we are already doing so. In the early stages, the circulation of rain clouds and wind will be shifted northwards, so that Southern Europe and the U.S.A could become like a desert. Later, agricultural patterns would be disrupted, and the rising atmospheric temperature will begin to melt the polar ice caps and glaciers. This would lead to a rise in water levels. If the ice caps melted completely, the sea level would rise over 50 metres, affecting most coastal cities. At some point , carbon dioxide would be released from solution in the oceans and from chemical combination as carbonate rocks. At this point a runaway vicious circle is set up, and the end result is the landscape seen by Venera 10. Of course, the ability of marine plankton to make up for the loss of the forests on land, and their own ability to cope with chemical pollution, are factors not yet fully computed, but it would be rash to rely upon them to bail us out.

In 1983 two reputable scientific reports from the Environmental Protection Agency (E.P.A.) in New York, and the National Research Council (N.R.C) indicate that the Earth's atmosphere is already 50% richer in carbon dioxide than in 1900, and that the average atmospheric temperature could rise by 4.5 C by 2040 A.D.-a change comparable with the change required to cause or end an Ice Age. If current trends continue, say the reports, weather patterns will be showing signs of changing in the indicated directions by the 1990's. By the Kyoto Summit in 1997, temperature rises had been scaled down somewhat, to 1.5 to 2 degrees, but increasing consensus on the need for action is emerging. The E.P.A in 1983 recommended a drastic reduction in the use of fossil fuels, and a major re-siting of human settlements away from coastal regions. To this must be added the deforestation of Amazonia; a third of the forest has disappeared from 1960-80, and the process continues. The E.P.A's recommendations are actually impracticable, for two reasons;-firstly, the improbability of halting the industrial revolution in Asia, and, secondly, much of the fertile land, not to mention fishing stocks, are only accessible in lowlands, which lie near river estuaries or coast lines! The exact timescale and degree of global warming remain to be established, but it will be on the agenda for the foreseeable future.

The second lesson from Venus is much more hopeful; the late Carl Sagan believed that it is possible to convert the conditions on Venus to a more Earth-like scene by seeding the atmosphere with a species of blue-green alga. This class of creatures contains some very hardy members which could survive and multiply in Venusian conditions, turning carbon dioxide into oxygen in the process. This would proceed, allowing the planet to cool, and steam to condense, leading to a more reasonable prospect over, perhaps, several centuries. The sulphuric and hydrochloric acids would reach the surface, and be chemically neutralized by the rocks, leading to salts and minerals useful to more conventional plants. Eventually , writes Berry, we would have a true sister planet. The recent discovery of sulphur metabolizing, anaerobic microbes living in superheated steam vents at over 300 C near volcanic vents on the ocean floor of the Mid-Atlantic Ridge should make matters far easier. The expanding science of genetic engineering should make some kind of hybrid between these thermophilic sulphur-eaters and blue-green algae quite feasible. A few tons of the appropriate organism, deposited into the Venusian atmosphere by rocket-launched balloons or landers, would be well within the technology of the 21st century. The lack of water in the Venusian atmosphere would be a major problem-but cometary ice could be mined stored and imported by large automated spacecraft with large inflatable storage tanks. Our astronomers could then sit back and wait for rain.

The exploration of Mars during this period resolved many mysteries. In 1965, Mariner 4 disposed of the Martian Canals which had been popular since the time of Schiaparelli in 1877. Mariner 9 orbited the planet in 1971/2 and photographed the entire surface. Huge mountains and canyons were seen, as well as sinewy valleys looking like dried up river beds carved by running water. Gigantic dust storms, and pole caps of water and carbon dioxide ice were observed. Viking landers finally reached the surface in 1976, and began the most daring extraterrestrial study to date-an analysis of Martian soil for evidence of bacteriological life. The results were inconclusive, but, with the passage of time, it seems most likely that no active life has been detected on Mars, and that the strange reactions of the soil were chemical. Soil sample returns, or a manned mission, are required to settle this question finally, although the biochemist James Lovelock points out that the Martian atmosphere, in marked contrast to Earth's, is in chemical equilibrium, and so, *a priori*, should be deemed sterile. Earth's atmosphere, in containing oxygen and methane together, gives its game away; normally, oxygen should burn methane out of existence in a very short time indeed, so that their co-existence proves that something is continually producing these gases over a very long time scale-and that "something" is Life. In Mars' early years, life could have evolved while there was running water, and been desiccated out of existence in the first billion years. However this summer (2000) it was announced that photography from the Mars Global Surveyor mission showed scattered gulley formations which point strongly towards water seepage in very recent times.

The temperatures are low, ranging from room temperature at the equator on a summer's day to well below Antarctic levels by night. The atmosphere is one third that found at the summit of Everest, while there is a small supply of water ice at the poles, and much more , it is suspected, under the ground as permafrost. It is now widely accepted that there has been liquid water, and even rivers and shallow oceans, in the past, and, it is believed, that there may have been long oscillations of climate between a "warm" Mars with rivers and an atmosphere, perhaps as dense as that in high Alpine regions, alternating with a cold, dry, airless Mars as at present. Even during the short Space Age, remarkable variations have been seen. The temperatures were dropped 30 C by a planet-wide dust storm in 1971 at the start of the Mariner 9 mission, while the Hubble Space Telescope photographed a colder, drier planet than was found in 1976. The long-term cycles described previously are believed from careful study to repeat at 20,000 year intervals. The Russians have made numerous visits to Mars with orbiters and landers, but have not enjoyed the success of their American counterparts. Phobos and Deimos, the two Martian moons, have been photographed by Russia and America, and have been shown to be typical of captured asteroid material, probably containing carbonaceous chondrite and hydrated material. Sadly, the Russian Phobos craft, in 1989, failed just as close encounter was imminent. America's Mars Observer, launched by Shuttle, and intended for a 2 year mapping and survey mission in 1992, failed on arrival at the Red planet; but 90 % of its work is to be re-attempted by a series of orbiters and landers to be launched over a 10 year period, which began in 1996. James Oberg (of N.A.S.A) has proposed the adaptation of Mars to human requirements, in a scheme employing an induced meteorite impact to melt the ice caps by heat of collision, and release carbon dioxide and oxygen locked up in the frozen soil. Underneath large polythene domes, plants could then be reared which would create an atmosphere and settle selected areas, whereupon human colonists could move in *en masse*. Obviously base camps of scientists and engineers could live in artificial habitats long before the climate provided an outdoor shirt sleeve environment! Gravity on Mars is 1/3 Earth's, so that the large scale structures needed for this plan would be easier to erect than on Earth. Inflatable domes, initially ferried from Earth, would be relatively light, and would be greatly extended by indigenous industry. Oxygen would be readily extracted from the soil, and used for life support and rocket fuel, as could the resources of the moons. One important discovery from Mars has been the dramatic influence of opaque dust on planetary surface temperatures, as demonstrated by the findings of Mariner 9; it was from this event that scientists turned their thoughts to Nuclear Winter, and, later, to impact induced climate change. Thus it was that the exploration of the Moon and planets has given us valuable insights-and, indeed, dire topical warnings-relevant to our lives and conduct on Earth; so much for the irrelevance of space exploration!

A Martian Century?

In 1996,with the launch of Mars Pathfinder and Mars Surveyor, a new phase of the Earth-Mars connection began. For 10 times less than the cost of the 1976 Viking landers, a lander-rover reached Mars on the 4th of July,1997, and started to explore Mars, watched from the Internet as it happened. 500 million hits were recorded at the Mars Pathfinder websites within a few weeks-a tribute both to the Internet's growing power, and to the enduring fascination of Mars for humans. These are the first of a planned 10 years series, including both Japan and ESA; all future Mars launch windows are likely to see a visit from Earth, culminating in a return sample mission by 2010-12; this will be, if all goes well, fuelled by propellents extracted by a previous Mars lander in situ from the previous launch window, and is aimed to ensure a cost of $400 millions, as opposed to the previously expected $2 billions. This fact, together with the work of Robert Zubrin on his Mars Direct proposals to send people to Mars on a similar basis will not be lost on NASA's Mars exploration planners. It is becoming possible to think of human missions to Mars, starting within 8-10 years of a political decision, at a cost comparable with the Apollo Moon landings-$25 billions or so. The aim would be to build up a permanent base there, with possible ultimate settlement during the coming century. Such a decision is ultimately political, and could be taken within 5-10 years; or, if Zubrin has his way, it could even be carried out by a consortium, outside the political arena. Thus, readers could well witness, or participate in, the dawn of the Martian century.

Most striking to those brought up before the Space Age will be the astonishing Pioneers 10 and 11, and Voyagers 1 and 2 of 1973-77, and 1979-2015?, respectively, in which the entire outer Solar system has been revealed, with the sole exception of Pluto-though even here, discoveries have been made from advanced Earth bound telescopes, and the Hubble Telescope. The Pioneers blazed the trail through the hitherto unknown asteroid belt, between Mars and Jupiter, reaching Jupiter in 1973. They detected intense radiation belts in the vicinity of Jupiter, as well as strong radio emissions from the turbulent atmosphere, and an internal heat production from the planet, which exceed the heat received from the Sun. Temperature profiles of the atmosphere revealed that a few dozen miles below the cloud deck were regions of 1 atmosphere pressure at room temperature. The composition, however, is hydrogen, helium, laced with methane, ammonia, and hydrocarbons, and is of interest because this is believed to resemble the primaeval atmosphere of Earth in perfect preservation, retained by Jupiter's massive gravity. The presence of helium 3 would be of enormous interest to nuclear fusion scientists, since it is thought that, with helium 3 and deuterium, fusion would be easier to achieve than with deuterium and hydrogen.

The Voyagers revealed a thin Jovian ring system, details of the cloud belts and weather, lightning flashes on a titanic scale, and the unexpected diversity of the four classical Galilean satellites, as well as finding several new smaller ones. The Icelandic Sagas of the 10th and 11th centuries described the daring voyages to Iceland, the land of Ice and Fire. Now the lands of ice and fire have taken on new shapes in the Galilean satellites. The four moons, Io, Europa, Ganymede, and Callisto-had-up to 1979 only been seen as small points of light in the largest telescopes, but now stood revealed as four distinct and exotic worlds. Io proved to be covered with active volcanoes with a sea of molten lava flowing over its surface. One volcano, Mount Pele, was observed hurling molten sulphur 400 kilometres above the surface. An immense electric current of ionized sulphur atoms was detected flowing between Io and Jupiter, carrying over 10 million million watts of current.

Europa was found to be covered with a sheet of water ice, with perhaps oceans of liquid water 50 kilometres deep between its surface. Arthur C.Clarke has suggested that, with the tidal heating caused by Jupiter's massive pull, these oceans might conceivably support some kind of life. Recent discoveries have shown that volcanically created hot spots on the ocean floors of Earth can support a bacteriological ecology entirely separate from the solar-driven one familiar to the rest of Earth's denizens; indeed, there are strains which can live and multiply at 300 C so that the idea of life on Europa is not unreasonable. On another note, Europa contains the largest supply of fresh water in the solar system, including the Earth, and, with only 1/6 Earth gravity, is eminently accessible. A Europa Orbiter mission for launch in the coming decade is being actively studied. These developments on Mars and Europa, as well as the discovery of an ancient fauna independent of photosynthesis on the floors of the Mid Ocean ridges around volcanic vents, have given increasing prominence to the discipline of Astrobiology-the study of the origins, evolution, and distribution of Life in the Universe. Indeed, Astrobiology is emerging as a major impetus and rationale for space studies and programmes.

Callisto and Ganymede, like most of the outer satellites of the solar system, showed water ice as well, hurled into solidified waves by meteorite bombardment at the same time as the Moon and inner planets. The period of planetesimal and heavy meteorite bombardment is thus confirmed as a solar system-wide phenomenon. In 1995 an extended follow up mission to Jupiter-Galileo-sent a probe into the dense atmosphere, relaying back data via its orbiting mother ship before an inevitable fiery end, while the aforesaid mother ship takes a two year winding tour through the moon's orbits. Galileo has passed by Europa at only 200 kilometres altitude, and shows clearly that the ice covered surface is extremely young, with evidence of flows and meltings like our own Arctic regions. An extended Galileo mission is now underway to explore these questions further

Moving outwards, the Voyagers have reached Saturn (1979-81), Uranus (1986), Neptune (1989), and, as of 1995, are some 8,000 billion kilometres from Earth, at the very edge of interstellar space. They are expected to be sending back information on the Solar wind well into the next century. Thousands of pictures of Saturn, its moons and rings, and weather systems were taken. The largest flash of lightning, 42,000 kilometres long, was seen at Saturn, as well as a retinue of new moons. Titan, the largest moon, was clearly shown to have a dense, opaque methane and nitrogen atmosphere, with the possibility of methane oceans, as well as solid methane continents! Recent evidence from the Hubble Space Telescope confirms the probability of a solid surface, which is now set for a visit by the joint N.A.S.A/ E.S.A mission-Cassini/Huyghens-launched in 1997. Like Galileo, Cassini is a second generation mission, with a lander and a long-term orbiter. The other moons-Mimas, Enceladus, Tethys, Dione, Rhea, Hyperion, Japetus, and Phoebe-were all photographed by Voyager, and should be revisited by Cassini.

In 1986 Voyager reached the large green methane rich Uranus, and imaged its newly-discovered ring system, as well as the extraordinary geology of Miranda, which can only be described as spectacular, with vast cliffs arising vertically out of chevron-shaped land masses. On a scale with Earth, these cliffs run up to 160 kilometres sheer.

Also in the 1980's and early 1990's the first close-up views of comets and asteroids began to appear-notably Giacchobini-Zinner in 1985, Halley in 1986, and Grigg-Skjellerup in 1992 among the comets, and the asteroids Gaspra and Ida have been imaged by Galileo en route to Jupiter. Within a few years, probes are expected to be sent to shadow, or land,on comets and asteroids. The Japanese are planning to return a sample from the asteroid Nereus with an ion engine powered mission in 2002-6. Although not as spectacular as the major planets, the asteroids are of immense industrial importance, since they are treasure troves of metals such as platinum, nickel and iridium, as well as water, and organic chemicals-and, moreover, can be reached from Earth orbit or the Moon with comparatively little expense of fuel. Indeed, some space industrialists favour the use of asteroid materials rather than the Moon in the next phases of major human activity in space.

Neptune was reached in 1989, and the second blue planet in the solar system stood out in lonely beauty against a stygian gloom. Its principal moon, Triton, revealed a strange landscape of solid nitrogen, with geysers of liquid nitrogen pouring onto a surface frozen to minus 230 C.

Space for the People

The next few years will see another remarkable development in the opening of the solar system to the public at large. Quite apart from the availability of superb CD-ROM images of the planets for use on home computers, a joint American-Russian private corporation is planning to land rovers on the Moon , carrying instrument packages for universities, prospectors, and the entertainment industry, for the provision of virtual reality images. N.A.S.A plans to set up interactive virtual reality stations on the Moon and Mars, while there is a proposal by this author to orbit an orbiting amateur 20cms. telescope to give students and members of the public, as well as science TV programmes, access to interactive, all weather astronomy. This project has been dubbed the Humble Space Telescope.

1996 saw the opening of a new era in unmanned planetary exploration, with the new class of discovery space vehicles. Under the programme slogan "faster, cheaper, better" , these missions are put together from conception to launch in 3 years, at a cost of $200 millions maximum including launch. The first, Near Earth Asteroid Rendezvous mission, visited asteroid Mathilde in 1997, and is now shadowing Eros 433 for a year, has achieved a spectacular first already; from a budget of $150 millions, NASA was able to hand back to the US government $38 millions! Later in the year, two spacecraft were launched to Mars, including the first lander in 20 years. With new technology, and the new paradigm, these missions cost $200 millions each including launch costs. As the Viking missions cost $1 billions allowing for inflation, the cost of unmanned space exploration is now reduced by nearly 10 times.

Despite the recent loss of Mrs Climate Orbiter and Mars Polar Lander (1999) the approach of Cheaper and Better missions is likely to continue in a modified form ; probably there will be a slow down from the "Faster" element and a better cash reserve to deal with contingencies and development problems. This is only the beginning. The privately funded Near Earth Asteroid Prospector mission is projected to set off for an asteroid in 2002, at a cost of $30 millions-five times less than the already cut price NEAR-and collect data for sale to the world's scientists, and industrialists.

In 1996, the UK's leading manufacturer of small satellites, Prof.Martin Sweeting of Small Satellite Technology Ltd, announced at the IAF Congress in Beijing, that he is aiming to build and launch a small lunar explorer/orbiter for $15 millions within 5 years, while the UK's Matra Marconi Space and Farnborough's DERA have tested and prepared for its first mission an electric ion engine, which could allow 10 fold reductions in the masses of interplanetary space craft. December 1996's Journal of the British Interplanetary Society carried designs of spacecraft of 250 kilograms, well within the scope of the privately run US Pegasus launcher, which flies for $10 millions a launch. These new technologies, the new management paradigms, and the emerging private cut price launcher market being built around the emerging mobile phone market promise to cut price desktop space exploration for science, education, and recreation for millions, built on missions costing $25-30 million dollars rather than hundreds as now.

Voyager is now the boundary of human enterprise, and has recently sent us back possibly the most remarkable picture of the century-Earth as seen from the borders of interstellar space-one pale blue dot against an immensity of stars, lit by the light of a bright star-the Sun. All of humanity-its wars, loves, fears, and dreams-take on their real significance against such a background. Indeed the only hope for a meaningful human future lies in the spreading of Mind beyond the tiny cradle of the Pale Blue Dot, so that our collective range becomes one with the immensity opening up before us. Should we fail in this, no mere Earthly creed will rescue us from this stark reminder of our parochial insignificance. I very much doubt that, in the longer run, even if we eked out a survival on our small beleaguered planet, our culture could maintain itself in the face of such a failure. Our very human nature would feel betrayed. Knowing what we now know, there is no turning back from the road to space-only the path to decline and eventual extinction.

Astronomy has also benefited from the opening up of the Infra-red, Ultra-violet, and X-Ray wavelengths. The origins of planetary systems, birth of stars, neutron stars, black holes have all been much better understood since the coming of space-borne observations, and the Hubble Space Telescope enables us for the first time, to record over the long-term events out in Space which previously have been fleeting snapshots. In the case of the planets, to better Hubble, we have to visit with powerful landers, or in the flesh. By 2010 the Hubble Space Telescope's successors and advancing ground based instruments are likely to be able to image extra solar planets, thus opening up a whole new field of exploration. With increasingly powerful radio telescopes, both on Earth and in Space, over coming years, it is quite possible that we shall know if extraterrestrial civilizations are really waiting to be detected. The long reach of Hubble into the further reaches of time and space is sure to throw more light on the origins and ultimate fate of the Universe, and to give a realistic setting to our eternal search for our place in it.

The period 1950-2000 will justly be remembered as the Age of the New Copernicus, and our ideas of the Universe and our role in it have taken a quantum leap forwards. It is now time to turn from the alien civilizations we hope to find over the coming decades to consider one extraterrestrial civilization in the making-our own.

Chapter 8

The Second Stage-Commercialization and Exploitation

The Space Age up to now has spawned two new industries-firstly, the transmission and processing of information-communications, meteorological, education, computer data, facsimiles, and entertainment-on a global scale, now amounting to tens of billions of dollars worth per year, and, secondly, the monitoring the Earth's ecological systems and resources with the multispectral scanners of satellites. These Earth Resources photographs and data from satellites are now for sale on the open market, and can be used for a variety of purposes, ranging from mapping to mineral location, crop management, and, in the case of Shuttle flown synthetic aperture radar, underground water location in drought-ridden Africa, or the discovery of lost cities in the Sahara. European S.P.O.T and Earth Resources Satellite (E.R.S.) imagery will soon be used to detect agricultural fraud in the European Union-at $6 billions per year, this would more than repay the investment! In 1997, selective crop spraying with fertilizer and pesticides was the subject of highly successful field trials by a group of UK arable farmers, using Spot imagery and global positional satellite (GPS) location data. More economic, as well as more ecological, use of these agricultural chemicals has been amply demonstrated by such targeting. These industries are likely to expand considerably in the next few years, with the eventual introduction of the satellite mobile phone and data transmission systems , and the accurate location of diminishing resources and pollution monitoring. Privately developed launcher systems in the United States should increase access to space, and reduce costs, for small to medium sized satellites. The air-borne Pegasus rocket, dropped from a Lockheed Tristar at 12,000 metres, fires its engines, and can take 350 kilograms , at a cost of $10 millions. The Pegasus is already in service, and several other companies are seeking to enter the market place in the next few years in the U.S, at no cost to the tax-payer. Newer launchers, such as the Pioneer Rocketplane Pathfinder, Roton single stage to orbit spacecraft, and Kelly Aerospace Astroliner, are being developed ,subject to capital financing, to serve the satphone launch market, and could lead to costs of $3000 per kilo to Low Earth Orbit in the case of the Pathfinder, or 1,200 kgs at an all up cost of $5 millions ; some of these ventures will fall by the wayside-but it is likely that at least one of these, or another, will succeed. Another market being studied is the rapid and flexible supply of the International Space Station by small carriers, for small but time critical cargoes. A study contract has been issued by N.A.S.A to four small private launch companies in September 2000 for exactly this task. Minisats of 250 kg or so could thus be launched in clusters of 4 at about $1.2 millions each. Small geostationary or lunar survey craft, coupled to the new ion electric engines could thus fall into the low tens of millions of dollars. This could lead to more privately funded technology in a variety of applications now priced out of court. Arianespace is the first and largest wholly commercially run space industry in the world, and is set to flourish. The exploration of the Moon by remote control , for scientific and cultural purposes, is becoming commercialized by the about the end of the century.

The early human space missions have been involved principally with exploration-of human physiology in the new environment. With the Space Shuttle, astronauts have also retrieved, repaired, and relaunched satellites and laboratories such as the European free-flying Eureka experiment, and N.A.S.A's Long Duration Exposure Facility. Checking equipment before discharging into higher orbits has also been a valuable role for astronauts. The manipulation of large satellites by humans in a weightless environment has demonstrated that astronauts can work productively there. The next 25 years will see increasing activity in space as a commercial and industrial base.

The U.S and the Russians have been moving steadily towards semi-routine space research. In the U.S., this is accomplished by the re-usable Space Shuttle and Spacelab, with extra work-space provided by the privately developed and rented Spacehab since 1993. The Russians, of course, have had routine access for 15 years with their Salyut and Mir space stations, automatic Progress supply ships, and Soyuz ferries. By 1985, the Americans had four Shuttle Orbiters-Columbia, Challenger, Discovery and Atlantis. These vehicles are exceedingly complicated and perform to very exacting standards. Not only do they endure lift-off and atmospheric re-entry, but they have to be refitted and repeat the journey many times. In the early 1980's, it was envisaged that by 1987 there would be 24 Shuttle missions per annum, rising to 30-40 by 1990, with a new fifth Orbiter in commission, shortly to be followed by Space Station Freedom. The early jobs for the Shuttle were the routine launching of all satellites from the payload bay, for which a stable of solid and liquid fuelled boosters were developed for onward transmission to higher orbits. Repair and maintenance of orbiting hardware was to be another commercial activity for the Shuttle; indeed, such activity has taken place with the Shuttle, but at a vastly reduced rate from what had been envisaged. It was also envisaged that during the 1980's, a range of new products, whose manufacture would be made possible by the routine access to the microgravity environment, would be in the global marketplace. The Space Shuttle would see the rapid change over from research to business in space based manufacturing. These hopes have proved premature, although much vital laboratory work has been carried out which points the way to several industrial processes which could create new markets when regular long term access to space, at a reasonable cost , becomes available. We shall see, firstly, the main merits and demerits of the Space Shuttle and Mir , secondly the processes and products which have been researched , and thirdly how the Shuttle's deficiencies are to be made good.

Firstly, the Shuttle has proven totally unable to perform at the level of usage originally intended. The Shuttle, even in 1985, the last year before the Challenger disaster , only managed 10 missions per year, and, even with 4 Orbiters in full swing, could at best have flown 12 times in a year. The refurbishment of the Orbiters, solid rocket boosters, fitting of the external tanks, and the need to reconfigure the entire payload bay between each mission, meant that turnaround time between launches could only be cut down at best to 59 days. This was better than writing off the vehicle for each trip and building another one, but hardly conducive to routine and reliable access to space. This fact, as well as the complexity of the machinery means that 3,000 people are involved in the maintenance and turn around of *each* Shuttle for each flight, or 12,000 for the whole system. These costs account for fully 77% of the cost of the Shuttle programme. In the beginning, the Orbiter had to land on the desert runways at Edwards Air Force base, 4,800 kilometres from its launch site. Now, weather permitting, they can at least land on a runway at the Cape. The solid rocket boosters, although reusable, have to be retrieved from the Atlantic by barge, and refurbished extensively each time-many authorities believe that liquid fuelled boosters, apart from being more powerful, would be safer, and easier to control and refurbish. The Shuttle fleet was intended to replace all America's launch systems ; this, it is now apparent, was beyond its capacity. The Shuttle, originally envisaged as simply a taxi service to an orbiting space station, was made too big and complicated because in the event, the politicians wanted it to do too many different tasks for as little expenditure as possible, and N.A.S.A, faced with the alternative of no space programme at all, had to accept a compromise from the politicians. The Challenger disaster, of course, led to an agonising re-appraisal. Firstly, it was accepted that the Shuttle could not do everything, and that a variety of expendable rockets was needed. This has now led to a flourishing stable of expendable rockets, with private corporations entering the field, to the general benefit of all. It has also allowed Arianespace to grow very successful in its chosen geosynchronous market.

Secondly, it is now accepted that the Space Shuttle system is an experimental system; although it has shown much more flexible and capable astronaut crews, and carried out several retrieval and salvage operations which could not have been done before, it cannot offer large-scale commercial industrial operations, nor long duration studies in space. Much valuable science has been done, and its enormous payload capacity has given the Shuttle the ability to take up large laboratories into space, and bring them back to Earth for refitting and reuse. A large and flexible tool chest can be taken up for repair work. Major satellites, such as Hubble, Solar Max, and the other Great Observatories, are designed for repair or upgrade facilities. However, many industrial processes in space need long duration exposure to microgravity-something the Space Shuttle cannot provide. The longest stay to date is 17 days, which is a bare minimum for crystal growth studies, let alone serious applications. The Russian Mir space station, by contrast, can offer missions of almost indefinite duration, but not in industrial quantities. The Soyuz can bring back 3 cosmonauts and 100 kilograms, while the newer Progress vehicles can send a capsule to Earth with 150 kilogram cargoes-enough for research purposes, but not for full-scale manufacturing. For serious space based industrial manufacturing, the Americans need a space station, while the Russians need a bigger logistics system. Everybody needs a cheaper, more reliable space vehicle.

The Space Shuttle system has been greatly improved since the Challenger disaster; the replacement Orbiter, Endeavour, is a much more advanced ship altogether. It can stay aloft for 18 days rather than 10 as previous models. It has a simpler thermal protection system for better maintenance. The original 1960's designed electronics in the fleet are now being replaced by 1990's style lighter, smaller, more reliable systems. The fuel turbopumps are being replaced by more durable pumps which can endure for 10 missions without servicing rather than 1 as before. The liquid-fuelled main engines are being improved, and will give slightly more power than the original. Perhaps the most positive aspect of all is that, in the wake of the Challenger disaster, it was decided to replace her with the newer, more capable Endeavour, and that the Shuttle programme came back to life in September 1988. Given an appalling disaster, which could easily have led to the abandonment of human space flight in America for a generation, the people nevertheless carried on. The Challenger disaster has probably set back the progress of humanity into space for a decade, but that, out of the ashes, a newer, firmer resolution has taken hold. A permanent space station will be essential for the next stages of industrial development and applications-a space transport system without an obvious destination cannot be sustained indefinitely, and secondly, it is evident that the Shuttle is an interim vehicle, and that a third generation spaceship will be required once the Shuttle has performed its main task-the assembly of the International Space Station.

From Research to Production

Space manufacturing rests on the idea that, in conditions of microgravity, certain industrial and chemical processes are made possible, or easier than under Earth's gravity. Over the last 25 years, many processes have been studied in metallurgy, crystal growth, protein separation, and electronics development, for example. Microgravity offers a hard vacuum, an inexpensive, inexhaustible, non-polluting energy source in solar energy, a heat sink to remove excessive heat generated in industrial processes, a vast storehouse of raw materials, easily accessible once in orbit, and isolated, quarantine conditions.

Microgravity allows many processes involving the separation of compounds of different densities, for example in bio-medical processing, perfect mixing also of substances of different densities and properties, such as alloys of light and heavy metals, or even ceramics, and the growth of pure crystals to proceed more effectively.

In medicine this is especially exciting. There are many compounds and biochemicals known to science which can have far-reaching effects in health and disease, but are impossible to obtain here on Earth in sufficient quantities or purity for use in clinical medicine. McDonnell-Douglas, in collaboration with Ortho-Pharmaceutical, are developing a process known as continuous flow electrophoresis. A mixture of dissolved biochemicals is fed into a chamber, and an electric field is run across the direction of flow. When the fluid reaches the pores sat the other end of the container, streams emerge in which the dissolved biochemicals are neatly separated out from each other. On Earth, this system does not work well because gravitational interference with the electric field reduces the precision of the system, and thermal currents cause mixing of the particles in the fluid-hence impurity. It is often impurities in biochemicals which render them unsafe for use because of potentially fatal allergic reactions, so that greater purity is a highly desirable feature. Several trials of this process have been flown on the Space Shuttle by astronaut Charles Walker. Improvements in yields and purity of several hundred fold over terrestrial results have been shown, and make this system a candidate for major manufacturing industry in space, once the logistical problems are solved.

These complex biochemicals are mostly proteins, whose molecular structures have a complex and characteristic distribution of electric charge on their surfaces; each would have unique flow characteristics when subjected to an electric current. The E.O.S (Electrophoretic Operations System) has already been used on several Shuttle flights to separate out two different proteins in solution. The results have shown a 500 fold gain in yield and a 5 fold increase in purity over similar process on Earth, and further improvements are expected. As of 1984, it was considered that there are 40 products for which the E.O.S has been considered a route to mass production. At first it was envisaged that a free-flying laboratory carrying out production with the E.O.S system would be launched into space from the Shuttle, visited and harvested twice a year by Shuttle astronauts on a routine basis, which regular supply of raw materials and servicing operations. With the Challenger disaster and subsequent reduction in Shuttle operations, this schedule has not been met, and it is likely that, pending the operation of the Space Station and next generation Space

Shuttle system, research will proceed, but no full-scale production will occur. It was estimated that the E.O.S. on a permanent space station would produce results at 1/3 the cost of a purely Shuttle run operation.

Perhaps the most exciting possibility is the tissue culture of pure beta cells. In 1929 the Canadian physicians Banting and Best discovered that an extract of Beta cells could treat diabetes by daily injections. Because these beta cells comprise only 2 % of the cells of the pancreas in humans, and are found in a few scattered islets (Insulae) in Latin, the new product was called Insulin. It transformed the prospects for millions of diabetics from certain death to a nearly normal life, and continues to do so. About 25 million people use daily insulin to stay alive; alas, the price is one or more daily injections, a strict diet, and the risk of several serious complications. These arise chiefly from subtle disturbances in fat metabolism which affect the smaller arteries , leading to blindness, coronary artery disease, strokes, painful leg cramps, gangrene, kidney disease, and sexual impotence. Good control does reduce these, but so far no human control mechanism yet devised equals the responsiveness of those islet Beta cells. Also, not every teenaged diabetic has the necessary discipline or aptitude to regulate the diet, exercise, and insulin dosages for proper control.

Pancreatic transplants have been attempted , but have proved unsatisfactory. Firstly, the pancreas is awkwardly situated behind the liver and duodenum and requires major surgery for access. Secondly, it is diffuse and poorly encapsulated-this makes it difficult to dissect out and lift up in one piece. Thirdly, the other function of the pancreas is to secrete digestive juices, which break down your food into simple chemicals for absorption by the body. With rough handling these juices to leak out of the pancreas, and digest it. Indeed, this can happen in some patients, resulting in an appallingly painful condition called pancreatitis. Finally, even if all these considerable surgical problems can be overcome, the recipient of this new pancreas promptly mounts an immunological rejection and is no better off. The fact that 98% of the bulk of the new pancreas is irrelevant to the patient's needs is no consolation, especially since the 2% which comprise the Beta cells are destroyed as well.

The tissue culture experiments offer a means of extracting beta cells from the rest of the pancreas with great purity and in high yield; The 8th Shuttle flight has already accomplished that. A small suspension of Beta cells could be injected onto a diabetic's liver by wide-bore needle, via a small tube inserted into a leg vein. This would not involve a general anaesthetic or surgery. A small, pure culture of foreign cells implanted into the liver will escape rejection, and function as normal, in 80 % of cases. It is also anticipated that this project could thus cure at least 10 million diabetics world-wide, and probably many more. For the beta cells, if they survive the transplantation as expected, will produce insulin naturally, in accordance with the bodily needs of the patient, and their hypodermic syringes can be thrown away, along with the diet sheets.

Apart from humanitarian considerations, this would save billions in hospital outpatients costs, operations for gangrene, invalidity from strokes and heart disease, and acute medical problems of excess or insufficient blood sugar levels. If a cure for diabetes were the sole result of 30 years of space exploration, the costs would be fully justified. McDonnell-Douglas expected that the eventual full-scale industrial operation of this one venture could eventually earn $15 billion in profits.

In late 1997, the first trials of beta cells cultured on the Space Shuttle have proved stronger and more capable of physiological functioning, in human volunteers, than Earth-grown equivalents, and the U.S Food and drugs administration have licensed space grown beta cells, suitably encapsulated against rejection, for clinical use in multi-centre trials in the summer of 1998. The likely endurance of a beta cell infusion is not clear, but indications are that 6-12 months between top-ups is achievable, with normal control of blood glucose. A great advance in the treatment of type insulin dependent diabetes seems to be on the way.

There are many other pharmaceutical products amenable to production in microgravity. An example is the enzyme urokinase, which has the ability to dissolve blood clots-it would, like streptokinase now in limited use for coronary cases, be a potent weapon in many diseases in which blood clotting is a factor, such as some strokes, lung clots and gangrene. These conditions could be greatly improved by rapid dissolution of blood clots. Urokinase and streptokinase can do this, but are fiendishly expensive-$200 for a millionth of a gram, and impossible to obtain in a safe degree of purity. Repeated use frequently causes severe allergic reactions. These severe allergic, or anaphylactic reactions can cause clinical shock or asthma, which can readily be fatal. This happens to those unfortunates whose lives are at risk from a single wasp or bee sting. Manufacture and purification of urokinase and streptokinase in space could reduce costs and give a safer product into the bargain. The anti-viral agent, interferon, recently used in alleviating multiple sclerosis relapses, is another candidate for the E.O.S, as well as Epidermal Growth Factor. This interesting substance is released in skin damaged by trauma, and promotes healing. The E.O.S has the potential for producing this substance in bulk and in great purity, halving the healing time of burns and injuries. Nasa Administrator Dan Goldin, in Turin in October 1997, expressed to this author great hopes for bio-medical advances from the Shuttle developed bio-reactor which has, among other things, made possible, for the first time, tissue cultures in 3-D. Important advances in embryology, and the study of the spread of cancer, could result from such work.

Growth hormones, anti-viral antibodies, and alpha chymotrypsin-a protective against some forms of emphysema-could all move from the laboratory into the market place as a result of space based experiments. The pharmaceutical industry stands to gain some $ 10 billions annually once there is routine access to space. Meanwhile, laboratory work on the Shuttle and Spacelab continues.

Recently it has been shown that in microgravity the central process of Genetic Modification-that of introducing altered DNA into a recipient cell with resulting incorporation into the hosts 's genome-takes place in microgravity 10 times more efficiently than on Earth.

The Salyut 7 space station has already been used for experiments in the manufacture of new, purer vaccines; impurities in vaccines have been implicated in severe allergic reactions both short-term and chronic, and increased purity will be an advantage. The purification and separation techniques essential in work with the delicate structure of genes may well prove much easier to carry out in space stations; there is also a built-in quarantine factor which could reassure a public, frightened by the possibility of mutant bacteria which could occur in gene manipulation experiments. A possible agricultural product could be a bacterium capable of fixing nitrogen directly from the air, and produce soil fertilizer in the process. This would be a great boon to poor farmers who would otherwise have to import expensive fertilizers.

Materials Science

In industry, microgravity and a hard vacuum make a powerful combination; the growth of large ultrapure crystals has already been proven. A good example is gallium arsenide, which can be used to make microprocessors and circuits which can work three times faster than silicon chips. Silicon itself can be prepared with greater purity and efficiency in space, and the doping and etching processes will proceed with even greater precision in the airless vacuum of space. The electronics industry is already the third largest after energy and the automotive industries; advancing technologies and activities in space can only accelerate this trend. Gallium arsenide crystals as large as rugby balls have already been grown successfully on the Salyut 7 space station-enough to meet the entire annual requirements of the then Soviet Union, and solar power panels made from gallium arsenide processed in space have appeared in the Russian space programme-perhaps the first example of a space programme enhanced by an industrial capability itself developed in space.

A further example of crystal growth is the supply of protein crystals for X ray diffraction analysis. Typically, the three-dimensional structure of a protein molecule is determined by the study of Xray diffraction patterns as a beam of Xrays is shone through a pure crystal of the protein. However, it is often difficult to obtain a sufficiently large perfect crystal of a protein under terrestrial conditions. In 1995, the 15.5 day Endeavour mission, conducted an experiment to grow a large crystal of urokinase. Urokinase is also secreted by malignant tumours, as a means of facilitating their spread through bodily tissues. If urokinase could be studied adequately in its three-dimensional structure, suitable blocking molecules could be developed, which would halt the spread of many cancers. However, its 3-D structure is poorly understood, owing to lack of an adequately sized crystal for analysis. On Endeavour, an attempt was made to grow a sufficiently large crystal to permit further analysis, and hence the design of blocking drugs for future cancer patients. Thus the access to microgravity promises greater understanding of chemical and physical processes, from which in turn new advances can be developed.

Perfect spheres can be made in microgravity, owing to the absence of gravity in the casting process. Two interesting products are perfectly spherical ball bearings of any desired size, and latex spheres. These latter are of use in measuring and monitoring devices for the control of the flow of gases and fluids, and are already supplied to industry from Shuttle based activities-being, in 1987, the first product specifically made in space for the market place! Ball bearings are essential in many load-bearing machines with moving parts, and to a great extent determine the working life of these machines. Imperfect ball bearings wear down more rapidly and shorten the life span by wear and tear. Ball bearings made in space promise more reliable machinery here on Earth, for example clutches and gearboxes. Life span increases of fivefold are considered possible particularly important in remote, poor, or inaccessible regions of the globe.

A further medical product may become possible from the fact that it is easier to create mixtures of different densities in space. Small glass spheres of 0.002 millimetres (5 microns) diameter with a minute iron core can be prepared reliably and at little cost. These spheres can be coated with antibodies raised against unwanted blood components such as leukaemic red cells, or perhaps HIV virus antigens, among others.. On infusion into the bloodstream, via an extracorporeal circulation, these artificial cells would latch onto their chosen targets. A large electromagnet can be switched on and used to trap all the glass cells. Their blood could thus be cleared of leukaemic, or other unwanted cells without recourse to dangerous marrow-destroying radiation or drugs. The Hammersmith Hospital in London has already experimented with the process of using glass "cells" in the treatment of leukaemia; however, the mass production of precisely engineered glass/iron spheres could become more economical with space based industry. More unexpected benefits are likely to accrue from this sort of technology.

On the Salyut 7 and Spacelab perfectly homogeneous alloys of indium and aluminium, and zinc with aluminium have been made. These will give metallurgists new alloys with unexpected properties.

From the foregoing, it can be seen that microgravity research and processing is already pointing the way to major advances in our understanding of industrial processes; some of this knowledge will help industrialists to make better products on Earth, while other experiments will lead directly to novel processes and products which can only be obtained by establishing a permanent, economical and reliable presence in space. At that time, a new industrial revolution will take off, in directions that may be as far beyond our sight as the uses of fire are to a chimpanzee. Many will, understandably perhaps, dismiss much of the foregoing as laboratory curiosities at best, and hype at worst. To them I can only relate the story of a curious metallic element, whitish in colour, which was discovered in the early 19th century, and forgotten as a mere curio, without any conceivable use, as it was well-nigh impossible to extract in sufficient quantities for any proper work to be done on it. 75 years later, electricity came into vogue, and at last the chemists had a chance to produce this metal in quantity. in A few decade later it leapt from the obscurity of a chemist's bench to take over the industrialized world. We know the humble lab curiosity as aluminium. The history of science and technology is full of "impossible" ideas which became practicable as a result of advances in materials science-and frequently the biggest consequences were totally unforeseen at the time. It is impossible to predict which major new products, inventions and discoveries will be made possible by the new materials processing techniques which routine access to space will give us-but, on any sober comparison with history, they will be enormous, and totally revolutionary. This arena is still in its infancy. This is a good time to mention a key law of space development which I call "the Law of the Golden Bootstrap". By this law, the revenues, and results of space technology will in part be ploughed back into aerospace, leading in turn to the increasing accessibility of space. For instance, it is not hard to imagine that advances in space derived electronics and the new materials , will be used to build improved space vehicles, which in turn will drive down the costs of space activities, and allow the creation of industries in space now deemed uneconomical. The commercial success of the Ariane 1 to 4 series of rockets has created a climate in which the development costs of Ariane 5 were justified commercially, which, in turn, allows for more ambitious space activities. Protagonists of major development rely heavily on the Law of the Golden Bootstrap to allow space development to become self-financing and effectively unstoppable. In retrospect, the hope that the Space Shuttle system was the vehicle to do this was premature, and a further stage of development is required. Meanwhile, much back-room, privately funded work into the means of achieving a permanent breakthrough, in strength, into the space frontier has been carried out.

The next crucial steps are, after a delay engendered by the Challenger disaster in America, and the collapse of the Communist Regime in Russia, at last beginning to take shape. The decision by the Americans to continue with the Space Station and Shuttle programmes despite Challenger, and the Russians to press on with an admittedly reduced programme despite their political and economic upheavals says much for the durability of the human enterprise in space. The agreement, in 1993, to carry out a series of joint rendezvous and docking missions leading to a merging of their space station programmes in the ISS Space Station in Nov 1998 should solve two problems in space development. These are, firstly, the issue of long-term access to microgravity for industrial and research purposes in the shape of a space station, and, secondly, the question of a stable political support. The American Congress has had a chequered history of support for the Space Station, as a result of which, repeated redesigns have led to a gross scaling down of the original Reagan proposal. This led Europe, and Japan, at various times, to consider going it alone, with reduced but independent stations. Principally because of Russian and European participation , the Republican dominated Congress has prevailed against those budget cutters who would have wished to axe the Space Station. Thus we have the irony that the next phase in the story is to be driven, not by rivalry, but collaboration. Dr. Gingrich, an ardent enthusiast for the new technologies, especially in private hands, proposed a scheme whereby N.A.S.A has a fixed budget for five years, rather than having to defend itself from financial cuts every year, with all the resulting detriment to long term planning.

The International Space Station , started in 1998/9, with a Russian base block consisting of a power generation unit, and station keeping motors. These first two units were launched and docked at the end of 1998, while the Russian habitation unit , long delayed, was rolled out on April 26, 1999, for launch in July 2000. The first crew are now expected to take up residence in late 2000. The Russian economic problems could delay construction, while NASA considers replacing some Russian contributions with US equivalents.

Meanwhile the Russian Mir station's fate has taken an unexpected turn. Although America wanted to see this facility de-orbited this year so that Russian funds could be concentrated on the ISS, there have been voices arguing that Mir, partially repaired, is a functioning asset which should not be wasted; European voices are suggesting that the US should not be allowed an effective monopoly in Space, while the private Space Frontier Foundation advocated the privatization of Mir as a possible first space hotel. An element of competition in space habitations is proposed as a stimulus for ingenuity.

This is undoubtedly an interesting proposition, but for success will require an increased and undisputed commitment by Europe to human space flight, and to building its own fleet of advanced space transport vehicles; FESTIP would need greater and speedier investment, in collaboration with Russia-or China, perhaps.

However, all this has changed; in early 2000, a Amsterdam based private consulting firm, MirCorp, has taken over the commercial development of Mir, offering its unique facilities , in the marketplace to scientists, industrialists, sponsors and tourists; In April 2000 the world's first commercially funded human space flight set out for mission of several weeks to re commission Mir , ready for commercially funded missions beginning in late 2000/early 2002. Indeed, the world's first fares-paying space tourist, an American billionaire, Mr Dennis Tito, is now about to embark on training for a 10 days tour in 2001. NMriCorp has announced plans to fly 2-3 missions per year and hopes to achieve permanent occupation of the station. Two new modules are being considered by Khrunichev for Mir in the next 2 years, as also are 2 for the ISS. Privatization of Space, it seems, is a growing activity in post Communist Russia! Will this all prosper? Many experts are doubtful in the extreme, but-time will tell.

With the successful launch of the Russian Zvezda module to the International Space Station, truss units, and 20 different laboratories and modules, are to be ferried up in 17 Shuttle loads, and assembled by astronauts with robot manipulating arms developed in Canada. A European Columbus module, and a Japanese laboratory module, are to be added, and the whole 430 ton structure is to ready for permanent occupation by rotating crews of four astronauts for stints of 6 months. Completion is expected in 2006. Since the Station is to be modular, and attached to a 100 metre long truss, it is probable that, in the first and second decades of the next century, it will be expanded as needs arise. Initially, its chief work will be biomedical research on the long-term effects of space missions, and the continuation and extension of the industrial and processing experiments I have described, leading in time to fully fledged manufacturing plants. In time, however, an improved ISS could be used to check out, fuel, and repair probes and crewed missions to other destinations. Satellites destined for geostationary or interplanetary missions could be checked out and mated to transfer vehicles in a garage section of the space station, allowing for less failures of vital projects. Orbital transfer vehicles which need never be launched from Earth, but fuelled in orbit, may be developed as the Space Station matures. Fuelling of vehicles in orbit automatically is an established Russian practice of some years standing. The powerful Centaur upper stage liquid fuelled rocket, which was grounded as a result of the Challenger disaster, could easily return in a new guise. Carried unfuelled into orbit, it could be "parked" in the garage, and fuelled from tanks in orbit. After transferring a payload to a higher orbit, the Orbital Transfer Vehicle, or O.T.V., could be brought back to ISS by "aerobraking", using minimum fuel. Boeing are planning to test fly their Solar Orbital Transfer Vehicle for comsat launches in 2004. The first uses of extraterrestrial raw materials will probably be supported and organized from ISS, for the supply of growing space industries and services. ISS, it should be remembered, is by far the largest and most long-lived of all international collaboration projects to date; it will involve all the nations of North America, Europe, and Japan-and now the Commonwealth of Independent States as well-in a 30+ year programme of intimate scientific, technological, and industrial collaboration on an unprecedented scale. More than the Space Frontier will be involved here. In building and operating International Space Station successfully, humanity will go a long way towards building a united planet. The Space Station will not, of itself, open the space frontier in a major way, for the benefit of all humanity. It has long been a tenet of space industrialization that the best bets for space industry were products of very small volume, and high rarity and added value-such as pharmaceuticals and electronic elements. There is one industry on Earth-the fourth largest of all-which depends on moving large numbers of very heavy items around and adding no value to them whatsoever in the process-the tourism industry.

Considerable market research has shown that, if access to space were reliable, proven, and no more costly than, say, a trip around the world by liner, as many as 60 million people per year would undertake the trip of a life-time-on a spaceship. This implies a potential revenue to the as yet unborn space tourism industry of as much as $300 billions per year. Space hotels, built out of standard space station modules, offering Earth views, astronomical perspectives, zero-gravity swimming and sports, including arm-powered flight, and health "spas" would not be so hard to construct once a new generation Shuttle is available.

Finally, it has been suggested that the industrial applications of the Space Station , while promising a few years ago, are now being matched by terrestrial technical developments-for example genetic engineering can replace many of the bio-processing techniques described earlier, and that, hence the scientific value of Man in Space is now greatly reduced.

This has some truth in a narrow scientific sense-but it should be remembered that the human development of space is in the final analysis not a scientific issue-it is a question of opening up an new frontier, indeed, an evolutionary niche. We have argued elsewhere that a fragile earthbound human civilization will not survive the impact of a major asteroid or comet, and that such an event is a certainty. This being so, dispersal of human civilization into Space is an essential insurance policy for our descendants-more than this, it is a supremely creative act, in that Life and Mind into the wider Universe. The construction of such facilities will be an immense and lengthy affair, and we have much to learn on the way; the ISS is several dozen times larger and more complex than our current space projects, while the habitats of the future will be similarly larger than the ISS. Thus, the criticism that the International Space Station's main role will be to teach humans how to live, build, and work in Space becomes, in the light of the threats to our terrestrial civilization and its evolution, not a criticism, but an endorsement; we will learn to live in Space because we must-it is what we are ultimately for!

Space for Everyman!

Recently, the first space tourist societies, offering suborbital "hops" of 80-120 kilometres altitude in re-usable rockets, are being organized; 16 groups have registered plans to fly passengers repeatedly into the upper atmosphere on a semi-regular basis within the next few years-spurred on by a $10 million dollar X-prize, to be awarded to the first person or group to achieve this goal without government help. Like the old Schneider aviation trophy in the 1920's, the X Prize's proponents expect that a barnstorming period of daredevil passenger space flight will open up, relying on the ingenuity of talented entrepreneurs and inventors. Even the prize's organizers have been amazed to receive 16 proposals before even assembling the $10 million purse!

The proposed future of space based materials processing, let alone the large-scale industrial development and settlement of space, is beyond the capacities of the Space Shuttle/ Space Station system, on grounds of cost, irregularity, and unreliability of access to space. In any event, by the year 2010, the Space Shuttle fleet will be ageing, with Columbia approaching its 30th birthday. At the time, ISS should be well established, with several industries dependent upon its work, and the return to the Moon well underway. The issue of replacing the Shuttle will have to be squarely faced within a few years at most in order to ensure a smooth handover of operations to its successor. For, if the Shuttle fleet's efficiency is seriously affected by ageing, Space Station operations could be forced to close down, and, in effect, a human retreat from the space frontier would be imminent. The other possible caveat on the development of ISS is the possibility of catastrophic change in Russia. Much depends on the ailing Mr.Boris Yeltsin, who will relinquish the Presidency in 2000. If Mr Vladimir Zhirinovsky, or one of his ilk, were to be President, all bets would be off. By this time, however, America and the other partners would have spent so much on the development of ISS that it would be politically and financially very difficult to terminate it. One acknowledged potential Russian leader, General Alexander Lebed, is reported as having a lifelong interest in matters aeronautical and astronautical, and would maintain or enhance Russia's space activities, economics permitting. However, the Yabloko Party is against all space activity. In the event, Russia's new President Mr Vladimir Putin has pronounced favourably on Russia's future space plans, stating that he was fully in favour of restoring and expanding Russian space activities as and when economic conditions allowed. On his recent State visit to Britain, he invited the British space programme to join in co-operation in this field. This author took the occasion to wish that Britain actually HAD a space programme. Some economists now believe that the Russian economy is over the worst of its post-communist transition, and is likely to improve now whoever holds the Kremlin. In the short/medium term, however, Russia's economic difficulties continue to raise questions as to her reliability in joint space ventures. Despite all this, President Yeltsin pledged increased funding (Sept 1997) for the Mir and subsequent space programmes, at a time when on board troubles were causing Western observers to write Mir off. Yeltsin pointed to a future in which Russian space activity would once again lead the world, and that "Russian girls would once again seek to marry a cosmonaut!" Developments in launcher technology mean that, even if ISS were cancelled, the ultimate opening up of the space frontier would be delayed only for a few years at most, and the main thrust of this book's arguments would merely be altered in timescale and detail.

There are, basically, three alternative scenarios. Firstly, for reasons of failure of nerve or vision, Mankind simply lets the space frontier slip out of reach , possibly for generations. This effectively means for ever, since Humanity's condition on Earth will begin to deteriorate so that our descendants, like the Easter Islanders before them, would simply not be able to muster the skills or political resolution to re-open the whole business. This very nearly happened after the Moon landings, but thankfully there was still enough residual momentum to get the Shuttle built and keep the dream alive. Secondly, the Space Shuttle fleet could simply be patched up, with gradual replacement by more developed successor vehicles, well on into the next century. There is considerable scope for improvement in the present Shuttle fleet, which, in a series of individually unspectacular steps, could result in a far better product. The motor car is still essentially the same beast after 100 years, but vastly improved. The Shuttle could have liquid-fuelled boosters instead of solids-this would confer greater power, throttleability, and safety. The rocket nozzles of the main engines could be replaced with new materials and shapes, allowing hotter, higher pressure reactions, hence higher "specific impulse". Maintenance checks and preparation procedures could be automated, allowing more rapid turnaround times and fewer man-hours on the ground. The external tank could be lifted into orbit, rather than dropped off in the Atlantic. This would provide a large hollow cylinder with enough tankage to build large habitation units, or fuel storage, or, at the lowest level, 35 tons of scrap aluminium for use in industry. All of these steps would improve the economics of Shuttle use, and in the absence of a reasonable third generation space vehicle, would suffice *faut de mieux*. The privatization of Shuttle operations starting in November 1996, is expected to lead in time to a halving of operational costs, and a doubling of flight rates. The modifications proposed would have to be grafted on to an existing set-up, and Orbiters would be continually in and out of dock while changes were being made. Alternatively, more Orbiters could be built-a fleet of 10 or 12 would give much more frequent and reliable access to orbit, perhaps one flight every 3-4 weeks. This would still not generate traffic volumes even remotely resembling the airlines, which the real opening of space for all Humanity requires. The third, and most desirable scenario, is the introduction of a third generation spaceship (one being Apollo, two, the Shuttle, and three, the new craft) before the Shuttle system finally reaches the end of its useful life. Even after the Shuttle is too elderly to be entrusted with human crews, it could still form the basis of a semi-reusable Heavy Lift cargo vehicle, with the bonus that, without the expensive and safety-critical Orbiter, the main engines plus boosters could put 100 tons into Low Earth Orbit at good rates. Fortunately, aerospace engineers have looked at this problem for some years, and are unanimous that a third generation Shuttle, which can meet all the problems of the current Shuttle system head on and overcome them, is feasible within 10 years or so. Both N.A.S.A and the Defense Department have issued study contracts for the design of a fully reusable single stage to orbit space vehicle with a minimum objective of reducing costs to orbit 10 fold, and costs of $200 per kilogram to low Earth orbit are considered possible-for a human, this amounts to some $14-$15, 000. Economies of scale make this likely to be bettered in time, especially as cheaper access to space means that more applications, now considered exotic, become practicable, in turn generating more players in the game, and reduced costs. Such a vehicle will require considerable potential traffic into space to justify its development costs, and increased industrial applications and tourism could provide such traffic, the more so as the development costs look much lower now than they did in the middle 1980's when such vehicles were first considered. It is a law of transportation, well seen in aviation, that the creation of a new vehicle and new routes soon creates the traffic to justify it. In the 1920's Papua New Guinea was inaccessible jungle, but the fledgling Australian airline opened up air routes, as these were the only feasible routes of transport in such a country. The entire social and economic fabric of Papua, and, come to that, Australia herself, is intimately entwined with the aircraft and its requirements; so much so that no rational Australian would contemplate a society without air travel. In the words of George Woodcock, of the National Space Society Board of Governors: "N.A.S.A is not in the business of meeting requirements, but of creating opportunities". In space development, both functions are equally vital. The key to the whole future in space, therefore, boils down to the third generation space-ship.

In order to achieve reductions in the cost of access to space, simplification of the launch vehicle is the prime requirement. Much consideration has therefore been given to the design of Single Stage to

Orbit (S.S.T.O) launchers, which can be loaded up like a bus or wagon with any desirable payload, rather than requiring individual redesign of the payload bay for each cargo, and the tiresomely complicated assembly of strap-on boosters and external tanks for each mission. Such advances, together with automatic assembly and count-down procedures, can greatly reduce the recurring labour costs in a space mission, which, account for fully three quarters of launch costs for the Shuttle. S.S.T.O.s are inherently safer and environmentally cleaner, in as much as they do not require the functioning of different stages, and their separation. Looking at the Rocket Equation there are three main routes to a S.S.T.O. Firstly, one can, using new materials, fabricate lighter structures and fuel tanks, so that the mass ratio between vehicle fuelled and vehicle plus payload is shifted to a higher value. An decrease in the ratio of to fuelled vehicle from 10 to 5% means a 1.4 times increase in exhaust velocity, hence in final travelling speed. Secondly, by raising the operating temperature and efficiency of the engine, one can increase the exhaust speed, which directly relates to travelling speed. The key here is, again, new materials in the fabrication of the engines and rocket nozzles-which is where some of the new materials processing might come in. Thirdly, one could reduce the amount of fuel mass by deriving some of the oxidant-which, in the case of a hydrogen/oxygen motor comprises 88 % of the total mass-(by drawing in some of the oxygen) from the atmosphere during the highly demanding first 15,000 metres of ascent. This requires a highly sophisticated engine, which can act as an air-breathing jet engine at low altitudes, convert to a supersonic ramjet at 2,000 to 6,000 kilometres per hour, and switch to onboard oxygen tanks, as a pure rocket engine, once the atmosphere has been passed. Such an engine could reduce the liquid oxygen requirements by 15-20%, but there is a necessary increase in engine mass, since it has to operate in three modes. This latter is the approach adopted by the British HOTOL/SKYLON. These vehicles would take off and land horizontally, using standard airport facilities; much remains to be learned about aerodynamics at the relatively unexplored regions of 8,000 to 25,000 kilometres per hour, which lies beyond the range of speeds encountered in the X-15 programme. However, much work has been done on supercomputers and wind tunnels, and sufficient clarification of the design has been done to prove that the concept is feasible, and, in the case of H.O.T.O.L, better than expected at the outset. The possible payload of a 230 ton fully fuelled space plane of the has been raised from 7 tons in Hotol to 12 tons in a 275 ton Skylon , and, in the preliminary materials research, the way has been pointed to new advanced light weight fire-resistant fabrics, before even an ounce of metal has been cut; this was enough to provoke a Government decision to cut off funding, and ensure that our competitors gain the benefit in due time. Tentative studies were initiated with the Russians for a possible aircraft launched version of H.O.T.O.L, using their massive Antonov transport plane. A Russian developed spaceplane, named M.A.K.S, was proposed, but has since been dropped. Alan Bond has formed Reaction Engines Ltd to develop the design further with industry backing from an Industrial Working group. As of late 1997, this group now requires $1.5 millions to verify the earlier proof of concept study, and $20 millions from the UK government to show confidence and convince foreign investors, who are ready to take on the project, under these circumstances. This, paid over 18 months, is all, according to Bond that will ever be required from the British taxpayer to bring Skylon into service. In Skylon, since most of the vehicle is fuel tankage, the balance-point is kept stable by keeping the heavy twin engines alongside the fuselage at the centre of gravity, rather than as in conventional rocket engines at the tail end of the fuselage. Advances in engine cooling technology has allowed reduction in the weight for cooling Skylon's engines from 18 tons to 600 kg. Unlike the US NASP, Skylon does not require a scramjet as part of a triple propulsion regime, and so is inherently simpler.

The preceding vehicles are all like aircraft, require a long runway, and an atmosphere, and depend critically on the solution to some severe problems in aerodynamics and materials science. As different approach was actively pursued by McDonnell-Douglas, as an offshoot of the "Star Wars" programme. In 1991, McDonnell-Douglas was given $60 millions to build a 1/3 scale prototype of a launch vehicle which could meet the Armed Forces' requirements to be able to place 9,000 kilogram elements of strategic missile detection and acquisition equipment into space on a routine, and cost-effective basis. This became redundant with the improved relations with Russia, but, in the astonishingly short time of 2 years, using new management methods and a totally new philosophy, the 1/3 scale test-bed for a future S.S.T.O was built, and undergoing test flights.The McDonnell-Douglas D.C-X vehicle was distinctly odd in appearance. It looked like a 14 meter tall truncated cone, rather like a huge pepper-pot, and takes off under rocket power, vertically. Unlike all previous vehicles, it landed vertically. The D.C-X used off the shelf Pratt and Whitney rocket engines similar to those powering the old Centaur upper stage; the fuel tanks were all in the broad lower end of the ship, giving stability for the payload on the top. The count-down was completely automatic, and the D.C-X could be set up for launch in 8 days by 35 people. The D.C-X, a small prototype was not intended to reach orbit, but to test the launch and automatic flying procedures. It has proved very robust. In 1994, it performed two launches inside a week, the second of which was marred by an explosion in a hydrogen fuel line, blowing our a 5 metre panel from the side. Nevertheless, it went into automatic landing mode, and reached the ground in controlled descent, with no further injury. As of 1995, it has been flown, by remote control, up to 2.5 kilometres, and tilted to 70 % from the vertical and back again, in a simulation of the manoeuvres it will have to make on re-entry. In the summer of 1995, it made flights of up to 6 kilometres, carried out the full transition from vertical to horizontal flight, and demonstrated rapid turn-around with minimal service and ground crew. It was then taken over by N.A.S.A, who replaced the fuel tanks and bulkheads with a new, ultralight lithium /aluminium alloy, developed in Russia, and, over the next 2 years were to improve the electronics and material construction, improving the flight "envelope", with a view to the next stage, the D.C-Y. This was to have been a full scale proto-type, designed to test the envelope up to orbital speeds. It would cost something like $ 2-3 billions to develop, over 3 years, hopefully starting in 1996 for a test flight in 1999. Two copies would be built, and full operations would be simulated. At that time, with new materials incorporated into the programme, if full orbital capacity were proved feasible, a fleet of full-scale operational spaceships would be built, at a development cost of $8-10 billions, by about 2005, ready to replace the ageing Shuttle fleet. This operational version was to be named the "Delta Clipper", and was, as McDonnell-Douglas say, intended to open up the commercialization of space in the same way as the D.C-3 (Dakota) aircraft did for aviation. Unfortunately, in July 1996, on its descent after 4 successful tests of the new modified renamed Clipper Graham, a landing leg failed to deploy, leading to a catastrophic crash.

Some of this work has been taken up by the Rotary Rocket Company, whose Roton spacecraft was rolled out in March 1999, with a view to orbital tests and activities within 2 years; this craft, like the DC-X, takes off and lands vertically, and achieves a good mass ratio with the use of light weight new materials. However it has two revolutionary features- firstly it lands, not with parachutes or retrorockets, but with deployable rotor blades rather like a sycamore seed, and secondly, to avoid complex and heavy turbo pumps the thrust from the main engine is distributed though 96 nozzles set onto a rotating disk at the base of the craft. Fuel is fed from the tanks to the engines by centrifugal force, thereby simplifying and lightening the structure. If this system works as planned, the cost of transport to orbit could fall ten fold. The craft is to be piloted from the start, thereby, it is hoped ,demonstrating safety and reliability.

Earlier that month, Vice-President Al Gore announced that the winner of the contract to build an X-33 third generation SSTO demonstration vehicle was to be, not the Delta Clipper, but Lockheed Martin's lifting body design. This is to use an advanced aerospike engine, liquid oxygen and hydrogen propellants, and was on course to fly 10 test flights to 80 kilometres altitude in 2000/1. However a rupture in the hydrogen fuel tanks has led to a long delay , probably until 2002 in the first flights. The aerospike engine, however, has been bench tested to full duration burns , while advanced thermal protection systems have been developed and validated. A plethora of X programmes-X 34, 37, 38 and 43-are developing and testing technologies and materials in the search for a more economical Shuttle replacement vehicle. Industry and NASA are to share development costs for the X-33 up to 2000's demonstrations. while the X Programme as a whole is to be upgraded over the next 5 years.

The current view is that, by 2005, enough work will have been done to allow an informed choice of technologies, with costs and risks limited, for a decision to be taken upon an operational Shuttle replacement vehicle. Another promising development is the advent of a totally private corporation, Kistler Aerospace Corporation, which has been founded in 1993 from privately raised capital. Walt Kistler is developing a series of fully re-usable rocket ships in order to open up the space frontier and its resources for the benefit of all humanity. These, like the Delta Clipper, are to be fully re-usable, robust, and launched with minimum fuss and manpower, at frequent intervals. To support operations, a Kistler Spaceport is to open in Woomera, Australia, in 1998. $250 million has already been raised for the first stage-the K1, which is to begin launching 2 ton payloads into orbit-uncrewed-in 2000/1, at $3500 per kilo, initially for the satphone market. A fleet of 6 K-1's is planned. Thus it can be seen that a fully privately developed next generation transport system in service by 2010 is a probability. Europe, for its part, is already at the early stages of its F.E.S.T.I.P programme, and is similarly weighing the merits of H.O.T.O.L, V.T.O.L, and other possible re-usable S.S.T.O's. The German Säenger is a two-stage spaceplane, but requires less advanced materials and engines for its development. There is a trade-off between high development costs and long lead times to a really advanced product, and a cheaper, faster approach to a more interim product. No doubt, taking the global space programmes into account, both paths will be followed. By 1998, the Festip study is settling on a technology demonstrator to explore hypersonic flight, to be built over 2000/4, leading to an Ariane 5 derived nearly single stage vehicle, carrying a small second stage plus payload. The reusable first stage could evolve into a hypersonic suborbital round the world rapid air transport replacement for Concorde by about 2015, and a full SSTO by about 2025.

There is a further aspect to Delta Clipper type vehicles which has been proposed. In achieving orbital velocity-8 kilometres per second-a "Clipper" is the ultimate spaceship in the following respect. The main consideration in planning any space voyage is not, primarily, distance, but change in velocity-DV in kilometres per second-needed to achieve a certain task. Thus, the DV required to get into Low Earth orbit is 8 kms/second-the largest, actually, we are likely to encounter for a long time. For transfer to escape velocity-e.g. for a lunar landing-DV is c. 3.2 kms/second, while for a lunar landing by rocket braking it is about 2.4 kms/second. The same value applies to lift-off from the Moon. This actually means that a Vertical Take Off/Landing (VTVL) single stage to orbit vehicle, having achieved orbit and docking with a space station, could be re-fuelled from depots and go on directly to the Moon, and land with an even bigger payload than it took up from Earth-there being 2.4 kms/second spare DV. Being designed as a rocket controlled vertical lander, it is already set up to function as a lunar lander, and would simply need to be refuelled on the Moon for the return journey to the Space Station, or even direct to Earth, since most of the braking for the return to Earth would be carried out by atmospheric friction, with the rockets only kicking in at the last stages. Looking even further ahead, a "Clipper", taken to lunar orbit and fully fuelled, could actually be sent to Mars, and landed on the surface, since the DV for escape-or landing-on Mars for an incoming vehicle is about 5.5 kms/second. The question of how much fuel needs to be carried up to orbit for fuelling up of the vehicle, as opposed to the cost of developing an optimized orbital transfer vehicle based on a Centaur liquid fuelled upper stage, will be a focus of intense debate in the coming decade. The Martian Delta Clipper, sent to Mars, uncrewed, in advance of a Mars expedition, would serve as a resident Shuttle craft ferrying people and material to and from the Martian surface to orbit, much as the Earth-bound system would act as a ferry between Earth and Alpha. Thus VTVL could have the potential to be the ultimate spaceship, operating between any planetary surface, while the travel between planets would take place in large cruise ships, evolved from space stations, in eccentric orbits, which would graze the Earth's orbit, swing out to Martian orbit, and back in to, Earth's orbit every 6 months. The U.S National Commission on Space Report, in the wake of the Challenger disaster, termed such craft "circling space ships", which, with the right initial orbit, would need minimal engine power on their cyclical trips between Mars and Earth orbits.

This scenario, of course, depends on one more industrial development, which should also come about during the next 20-25 years. The majority of high performance rocket engines are driven by liquid oxygen, and hydrogen. Of the total fuel mass in such a rocket, fully 88% is oxygen, only 12% being hydrogen. It has been shown that, once traffic between orbits-whether from Low Earth to higher /Geostationary orbit, or to lunar orbit-exceeds about 50 tons per year it becomes increasingly economical to extract, store and supply fuel from the Moon rather than ferrying it all from Earth. Indeed, oxygen can be supplied from the Moon at 5% the cost of shipment from Earth, because of the difference in DV. Energy required-hence fuel and cost-increases as the square of the difference in speeds, by the formula $E=1/2\ mv2$, the classical formula for momentum, where E is energy, m is mass, and v is velocity in appropriate units per second. Lunar regolith is 40% oxygen in the form of oxides, so that there is no shortage! Hydrogen, being only 12% of the fuel mass, will presumably be supplied from Earth by bulk transport in the earlier stages of extraterrestrial development, but, in time, should be readily available either from hydrolysis of water from the Martian poles, or comets and asteroids as appropriate, or from the solar wind, whose charged particles-protons and electrons-are the building blocks of hydrogen. This could be harvested either in high orbit, or from the regolith, which receives a healthy pelting from the solar wind over time. Powdered aluminium, mined from the Moon, would provide a home grown substitute for hydrogen, and, together with lunar oxygen (Lunox) could cut the cost of supply of Earth orbital facilities 10 fold. Studies in America and Japan on regimes for mining oxygen, silicon, and metals from the lunar surface are proceeding apace. The discovery by the lunar orbiter Clementine in 1994 of evidence for water ice in the south polar Aitken basin-and strengthened by Lunar Prospector in 1998-is likely to greatly simplify the task of establishing a permanent base on the Moon.

Major New Industries

By the year 2020, a substantial Earth-orbital cum lunar industrial base could be in place, built on economical and regular access to Earth orbit, a space station complex with established manufacture of high-value products and processing, scientific and technological research, and the first use of extraterrestrial products-lunar/asteroid derived rocket fuel. The next steps would seem clear enough. Firstly, there will, on all available market research evidence, be the beginnings of a substantial tourist industry, to and from Earth Orbit, with the first orbital hotel complexes, based on the ubiquitous space station modules, well underway. Newt Gingrich told the annual National Space Development Conference in 1995 that he expects tourism to be a major generator of profits in Space by 2015. Japan's Shimizu Corporation thinks the same way. As the fourth largest industry on Earth, the tourist industry, if it gains access to space , should be a major generator of space development, and, furthermore, with the active consent and involvement of people who nowadays view space as marginal, at best, to their lives. Certainly, a vast growth in space-plane traffic and manufacture will result, generating millions of jobs in a world otherwise painfully short of new major growth arenas. No-one could possibly imagine, in 1945, that 30-40 years later, 800 million people per year would cross the Atlantic by air, at prices, in real terms lower than a Channel crossing in 1945.

Having established the extraction of oxygen from extraterrestrial surfaces for fuel, and probably life support purposes as well, and an industrial basis by this period, attention will then turn to the wide-ranging use of lunar resources for much larger industrial activity, and the beginnings of the third stage in expansion-Settlement. Setting the scene, however, we can point to two facts. The lunar surface contains basalt, iron, aluminium, silicon and titanium, among other things. Research into the economic mining of these materials for the fabrication of very large structures has already been conducted by private research institutions, such as the Space Studies Institute,. Also, work on the production of glass, fibreglass, and concrete from lunar topsoil is well established. Organic chemicals, hydrated minerals, and nitrogenous material is well represented in the asteroid population, which has members which graze Earth's orbit. These stray asteroids can come within a 3 weeks space mission from Earth, and with a DV requirement of only 1.5-2 kms/second-less than for a lunar landing. It is far more constructive to use these Near Earth Objects for our own development than simply to wait for an inevitable annihilating impact, or to risk multiplying the hazard with 50 megaton nuclear explosions, which would require a new class of heavy lift vehicles for their launch. If civilization is not to be wiped out, something has to be done with the near Earth asteroids, of which half a million are thought to exceed 500 metres in diameter. Why not turn them to our account?

In summary, then, there is no space industry at present conceivable by humanity whose principal raw materials cannot be obtained more cheaply from space than from Earth, given the technical and economic developments envisaged by the year 2020-excepting, of course, the human beings themselves!.

Another invention , the mass driver, makes the transportation or raw materials extracted from an asteroidal or the lunar surface into high orbit for manufacturing purposes even more attractive. The electromagnetic

Mass Driver first saw the light of day in Arthur C. Clarke's "A Fall of Moondust", in the early 1950's. This was a several kms. long rail, fed by electromagnets, along which a vehicle could be propelled to escape velocity by nuclear generated electricity without recourse to rocket fuel or engines; it neatly sidesteps the rocket equation by having the motive energy supplied by a fixed power plant on the ground, similar to a linear induction motor as used in some advanced rail systems. In 1969-74, the late Professor of Physics at Princeton University, Dr. Gerard O'Neill proposed the use of a Mass Driver as the preferred method of transporting large amounts of raw material into high orbit for large-scale manufacturing. At first, this looked fairly starry-eyed, since Clarke's version could only achieve 30 g's acceleration, which meant that to achieve escape velocity from the Moon, the putative driver would need to be 9-10 kms long-a gigantic structure. Dr. O'Neill and Dr. Henry Kolm began laboratory work on a bench scale mass driver, and were able, over the next 8 years, to demonstrate the Mark 3 Mass Driver at their Conference on Space Manufacturing. In 1982, a 30 cms long section of mass driver, with induction coil and bucket, accelerated to 500 kms/hour in 0.007 of a second, over 30 centimetres. This translates to 1,800 gravities acceleration, or lunar escape velocity over 150 metres. A fully fledged mass driver of this power would weigh 75 tons, or 3 Shuttle loads, while the supplying solar power unit would require 15 tons, if not built indigenously. Using a mass driver, 100 grams of lunar material every second would be accelerated in buckets along a series of electromagnetic induction coils into lunar orbit; the buckets would be decelerated after release of their contents, shunted off into a return loop, and used again. In high lunar orbit, the lunar material would be caught and processed into appropriate forms for construction or manufacturing purposes. 8-10,000 tons of material per year could be transported with one mass driver, at a cost of 1-2 pence per kilogram. A bonus with the mass driver is that, attached to a spacecraft or small asteroid, the ejection of matter by solar derived electro-magnetic induction provides a rocket which can run on any propellant, since it is mass itself that provides the thrust, not chemical reactions or heat. Thus, lunar slag, inert gases, or asteroidal rock, if broken into 100 gram packets, becomes a propulsion "fluid", and could give a Specific Impulse comparable with a high performance rocket engine. Thus, for freight transport between the planets, asteroid belt, or even *of* asteroids, we have a new, rugged, cheap and universal workhorse in waiting.

Robert Waldron of Rockwell International has already produced a working prototype which will process 100 times its own weight of lunar material per annum. His research into lunar reaction chemistry was essentially complete in 1983, and has thus proven practicable a vital plank of Dr. O'Neill's scheme. This work, like the work on the mass driver, was financed by an organization called the Space Studies Institute. The European Space Agency has studied the use of solar concentrators to distil desired elements from the lunar dust, while Japan's Shimizu Corporation is employing a team which, under Kenji Takagi, has developed an interlocking regime of oxygen and mineral extraction, rocket fuel production, and radiation shielding by slag production, which could allow a Japanese lunar base in about 2020, in four stages, leading to a 15 person outpost over the following generation

Owing to the absence of gravity, very large-scale structures can be assembled in space without regard to aerodynamic or structural considerations. For instance, very large mirrors, or cylindrical containers, or scaffolding are limited on Earth because they would fall apart under their own weight, be impossible to transport, let alone launch into orbit, or would face unacceptable damage by wind or rain. In space, ultralight gossamer threads of material can be extended for kilometres if need be without fear of unsoundness; nor would they require decades to transport. Component parts of great size can be moved with little DV and few men-we have already seen this in the acquisition of large satellites by one or two pairs of hands in recent years! Waste heat and material can be easily disposed of in the limitless vacuum of space, without threat to delicate ecosystems. Recycling of water, and oxygen through plants on board space stations and hotels, and other habitats has already been pioneered on the Russian space stations, and would be second nature to the space economy, where conservation of scarce resources would have a built-in urgency not evident on Earth. Even without radical new technologies in space, it is likely that a change in cultural patterns would slowly emerge from these developments.

We have now outlined the growth of industry to large scale operations, and seen that tourism, Earth's fourth largest industry, could become the first bulk industry to be transferred to space. The third largest industry, energy generation and supply, will see application direct to Earth of the new frontier, and is the other half of the resources we can expect to gain from space development. With the infrastructure I have sketched, energy production and generation from space becomes the next phase of industrial development.

In an earlier chapter we have looked at the looming problems of safe, sustainable energy supply and generation to a growing population with legitimate aspirations for a more ample way of life; in the process, we have found considerable problems with all the conventional options-.fossil fuels all contribute to global warming and are, moreover, on a historical time scale, non-renewable. Nuclear power is not acceptable in bulk because of the unresolved problems of waste disposal, not to mention diversion of material into criminal hands. Uranium is, in any case, a non-renewable scarce resource. There is also the problem of safety-Chernobyl springs to mind. Fusion is not likely to be available in sufficient quantities, nor at acceptable prices before 30-40 years, while its likely cost is much greater than the next stage of space industry. In considering the provision of energy from space, we must bear in mind that it is envisaged as being the natural outcome of an infrastructure as described above, which will already established; so that the cost of the following proposals will be much less than if considered from the present perspective. No-one with the perspective of 1930-40 aircraft technology considered mass passenger air travel a realistic option.

Looked at dispassionately, the logical, natural and almost indefinitely renewable energy source for the next century is solar power; the Sun has about 6 billion years of useful life ahead, is available without producing greenhouse gases, and has neatly solved the problems of nuclear fusion-at a safe distance of 150 million kilometres. It has some disadvantages as at present constituted. Firstly, it is weather dependent-there is precious little to be had through a thundercloud, let alone night time. Secondly, it is latitude dependent, which means that northern regions have much less access than remote deserts and jungles. Thirdly, atmospheric attenuation means that, by the time it reaches the surface of the Earth, it is very spread out, so that to collect useful amounts would need very large areas to be covered with solar arrays-for America, something like 8% of the entire land area of the United States would have to be given over to solar power collection.

There is already a considerable body of experience in the generation and utilization of solar/electric power in space-the familiar "wing" on most spacecraft, including the Russian Mir station-are none other than solar power panels; so that the following proposals seem to follow logically from current practice.

Peter Glaser, Chairman of the Arthur D. Little Power Company proposed in 1969 a radical solution to the world's energy crisis based on space-based solar energy. He envisaged huge solar power panel arrays which could collect 10 billion watts of energy each and convert it to microwaves. These microwaves would penetrate the atmosphere and clouds without hindrance, and reach huge collecting areas placed in desert regions. There the microwaves would be converted into electricity, and supplied to the grid. They would be independent of weather, consume no fuel, and be almost everlasting. Conversion of light to electricity via microwaves can be done with 80% efficiency, rather than 16% as in direct photoelectric conversion, and the initial collection of light at 8 times terrestrial intensity permits a correspondingly smaller solar collector in space than on the ground. Multiplying the two factors, it can be seen that a space-based system would occupy 40 times less land area than a terrestrial one-surely a major attraction in a time of growing populations and fierce competition for land. 10,000 megawatts is about equal to 10 Sizewell B's, or 10 Three Mile Islands, and, it was estimated many years ago, would cost $35 billions to build in space, assuming all components to be lifted from Earth. Much work has been done on improving the efficiency of solar energy to microwave conversion, and in design of the solar arrays; as we have seen, the economic and logistical environment of the 2020 period, in which this programme could start, is much clearer, and more favourable than when this subject first appeared in the early 1970's. To achieve similar results by nuclear fission would cost about $30-40 billions in the present climate, with all the environmental worries attached. It was envisaged originally that a 10 billion watt solar power satellite would weigh some 80,000 tons and comprise an array of silicon solar power collectors some 20 by 5 kilometres in size. Research has since increased the efficiency of solar power collection, conversion to microwaves, and transmission equipment to the point that the above weight can be considerably reduced, and it is likely that experimental and demonstration models will be much smaller. Tests on the ground (as of 1980) have already shown that solar energy can be converted into microwaves, transmitted, and re-converted into electricity with acceptable efficiency and no detectable threat to bird life. Ground based transmissions of 100 kilowatts have already tested and proven this. In 1995, at the Space Studies Institute sponsored Conference on Space Manufacturing, beamed microwave power was used to maintain a model helicopter in controlled flight before a delighted audience! The levels of microwave radiation are approximately equivalent to that received by a cook from a microwave oven with closed doors. It is therefore clear, as of 1999, that there are no technical or ecological barriers to the solar power satellite economy; however, with the present glut of oil interest in the concept has waned, and American government officials rejected proposals for their construction in the early 1980's on the grounds that to launch one from Earth would be too costly. However, the schemes proposed by Dr. O'Neill and Peter Glaser do not visualize a launch from Earth in the first place, but consider that the solar power satellite economy will be a natural outcome of the space industrialization scenario we have outlined in these pages. We have seen that, by about 2010, there should be in place a thriving space-based industrial system based on transfer and collection of information, servicing, materials processing, and probably tourism, supported by lunar raw materials and a lunar base. This will be at a time when many reserves of fossil fuel will be running down, and rapidly developing Asian nations will be placing legitimate and heavy demands on these resources. Many strategic metals will also be scarcer, and the current generation of nuclear plants will be approaching decommissioning-with no great enthusiasm for their replacement! To cap it all, the world's population will be approaching the 7 billion mark. It is against this background that the world's leaders will find themselves in desperate need of an acceptable, practicable, large-scale source of energy and materials. At this time the Earth/Lunar/ solar power satellite economy will be the only solution which can meet these requirements, for the space-based industries required for the job should be in place and ready to begin work. Solar power satellites, meanwhile, have been pronounced safe and capable, and, subject to the availability of space industry, their costs have been reduced with each successive study. A new study, fuelled by a $150 millions Government grant to NASA, is a rumoured imminent result of the 1997 Kyoto summit on global warming. Dr. Gerard O'Neill, in his best selling book, "The High Frontier" outlined the growth of space based industries to the point where S.P.S (Solar Power Satellites) can be built with lunar resources rather than terrestrial ones. As of now, it is possible in some designs to build a S.P.S with only 5 % input from the Earth, thus reducing the cost 20 fold, once the infrastructure I have described is in place. Dr.O'Neill envisaged a manufacturing colony, or outpost, of engineers, construction workers, and scientists, with supporting staff, where lunar material would be processed into S.P.S and other large-scale construction projects and shipped off to high Earth orbit. Such an outpost would begin as a "seed colony"-an extension of the Space

Station-of several dozen astronauts, to be expanded into a near-normal habitat of 5-10, 000 inhabitants, spending tours of duty of 2-3 years, earning their upkeep and pay by means of manufactured goods and S.P.S which would derive revenues-tax free, if Newt Gingrich has his way-in the markets of the world. The evolution and conditions of space colonies will be discussed in the next chapter, but this first outpost colony, named Island One by Dr. O'Neill, has been under active consideration by N.A.S.A at the Ames Research Centre in 1977, as a response to the "High Frontier". They came up with a detailed design for a habitat with some 10,000 inhabitants. In view of the physiological findings outlined earlier, "artificial gravity" will. of course be provided for the health of its inhabitants. In the "High Frontier", Dr.O'Neill envisaged a start on Island One before the turn of the century, but the Challenger disaster, and the initial over-estimation of the Shuttle's capabilities preclude this. I would think that a timeframe for starting serious design work, with the beginning of construction shortly thereafter, could be 2020-30, especially in view of the energy market and the ecology of the Earth which seems likely to obtain at that time. Much depends on the speed with which regular, and cheap, access to space can be obtained, which in turn hinges on the next-generation Shuttle. Further factors working towards "Island One" are the growing internationalization of space activity, spreading development costs over a broader base, and the advent of private entrepreneurial activity on the other, both in the launcher and applications markets. In 1991, Rep. George Brown, a Democrat, moved-and passed-a Bill in the U.S. Senate committing the United States to the long-term goal of leading the colonization and settlement of the solar system. Means and times were wisely not specified, but the transition of Gerard O'Neill's concepts from a class exercise in 1969 to official U.S. policy inside a generation is surely a remarkable achievement.

It is quite conceivable that a privately funded consortium of multinational mining, pharmaceutical, electronics, and aerospace firms could collaborate on the construction and operation of Island One during the next generation, and Dr.O'Neill's ideas are attracting the interest of industrialists and business men as well as space enthusiasts. For Dr. O'Neill was no mere visionary, but an experienced scientist who launched an actively on-going plan aimed at realizing these ideas , to be described later, which has survived his death. The large Island One colony would be built from a smaller seed colony using lunar derived materials, from a lunar base crewed by 20-30 residents. In considering S.P.S, we have considered the "classical" O'Neill route to Island One, which, in providing Earth type habitation for long-stay crews in large numbers, is the bridge between space industrialization-Stage 2 in my scheme-and full-blown settlement and colonization, which is Stage 3. Since O'Neill launched this whole movement to colonization, it has become possible to list three possible paths to Island One, or something similar; the first is energy production, in the shape of solar power satellites; the second would be a development from tourism, to hotels and thence to whole resort complexes, as has happened on Earth; the third is via long duration interplanetary travel and development. If the concept of "Cycling Spaceships" to Mars, on a permanent ellipse between Earth and Mars with rendezvous with Delta Clipper class Shuttles at planetary contacts is chosen, it is natural to imagine that the cycling spaceship component of the journey, will come to resemble Island One. This would be still more the case if research voyages of, perhaps, several years' duration were to take scientific crews-like Captain Cook's voyages, or the H.M.S Beagle-to remote destinations like the outer planets.

Where is the first Island One to be situated? To answer this, the requirements must first be stated. Island One will be a large structure so that, it must not be subject to the atmospheric drag of the Earth. The space station Skylab, launched in 1973, orbited the Earth at 500 kilometres, and re-entered in 1979 because at that altitude, there is still a very rarified atmosphere. At only one millionth the pressure of air at Earth's surface, it was still enough to retard the orbital velocity of the large Skylab, so that a fiery re-entry was inevitable; its timing was also hastened by the effect of the solar wind on the upper atmosphere. It could only have been prevented by a boost from a rocket engine, and Skylab was not fitted with one. The Russian space stations are, and must be boosted weekly to remain in orbit. For a structure the size of Island One, this would be totally unacceptable, and a re-entry would be catastrophic. Island One will be comparable in mass to the Great Pyramid, and so will need to be several thousands of kilometres from the Earth as a minimum.

As it is to be built from lunar materials, and to be used for the creation of space industries derived from lunar materials, Island One should be reasonably accessible from the Moon as well. It should, in fact, be in gravitational balance between the Earth's, Moon's, and Sun's gravitational fields. In 1770 the French mathematician, Joseph Lagrange, calculated that there were five regions of equilibrium, or libration, between the three gravitational regions. Lagrange points 1 and 2, or L1 and L2, are situated in front of and behind the Moon, as seen from Earth, respectively, at lunar-synchronous orbit-so as to remain above the same point of the lunar surface. L 3 is behind the Earth, as seen from the Moon, whereas L4 and L5 mark the apices of two equilateral triangles whose baseline joins Earth and Moon, thus L4 and L5 are each 400,000 kilometres from Earth and Moon. The recent possible discovery by the spaceprobes Clementine and Lunar Prospector of water ice at the South Pole of the Moon, renders the O'Neill scenario far more viable than even the keenest of his early supporters could have hoped for.

L5 is the favoured site of the Island One colony, and for the subsequent beginnings of colonization. L2 is significant for the collection of lunar materials. L5 is a region, rather than a point, covering a volume of space about 130,000 kilometres wide and deep. This affords considerable room for human occupation!

So it is, then, that, in the next generation or two, space-based industry will grow until it requires the use of extra terrestrial resources, for its own advancement at first, but later to contribute to the solution of those problems resulting from successful industrialization of the Earth. Island One, or something similar, seems a logical step in the development of space resources for an advanced technological civilization. It should also be seen, however, not as purely a means of meeting real and urgent requirements.

For it is my belief, based on the study of history and evolution, and the work of men such as Gerard O'Neill, Konstantin Tsiolkovsky, Wernher von Braun, and many others, that this step is also the road to new opportunities of gigantic scope, for the benefit, ultimately, of all peoples of the Earth. Island One seems to me to be not merely an endpoint, but a beginning of a new phase in the history of Life.

Chapter 9

The Third Stage-Settlements.

In the course of the 21st Century, space developments could lead to the construction of Island One, a large, self-supporting space habitat, which will act as the focus of an economic nexus linking the Earth, Moon, and high orbital industries. It will also be the jumping off point for the establishment of very large scale manufacturing and construction facilities. In providing near-Earth conditions for workers of many types for long-stay tours of duty, Island One will justly be seen as the first true space colony. It is true that the original O'Neill concept of Island One as a city of 5-10,000 inhabitants will, owing to advances in design and robotics, telepresence, and the like, not be as large as this; nearly 20 years' work by O'Neill and his associates have been largely directed towards making the initial requirements for the Island One/ Lunar base / solar power satellites economy as small as possible, so that it could be done within Shuttle technology, in a reasonable time frame. A next generation shuttle will make the whole concept more realizable, and, being human, most space advocates want to see their ideas on course during their life-times! The point of Island One is that it should be the smallest feasible self-supporting industrial base capable of carrying out the long-term expansion of human activity in space, with, minimal dependence on Earth, and the vagaries of its politicians and public opinion. It should thus be capable of providing its own food, air, water, raw materials for construction, and labour. It should also provide home to a sufficiently large population to allow breeding of a healthy, representative human population. All this should be possible using lunar and asteroidal resources after initial set-up with workers from Earth, and require only the input of small amounts of sophisticated products-such as electronic components, specialists, and so forth, from Earth. At this stage the unlimited expansion of space related activity would bear very lightly on the average person's tax bills, since the enterprise will be self-supporting and financing. For Island One will have several functions-solar power satellite fabrication, large tourist resort and hotel provision and construction, a "Noah's Ark" in the event of a civilization-threatening catastrophe, the development of cycling spaceships for interplanetary travel, and the production of more "Island Ones", or, with the benefit of experience and larger-scale space industry, bigger and better Islands. Islands Two and Three will be true space colonies, and built in large numbers. Thus Island One, whatever its initial size, is the nodal point between space industry and habitations-for it is from Island One that the real "Conquest of Space" will begin.

It is worth while looking at the origin of O'Neill's ideas and how they have come to achieve the prospects of reality. Professor Gerard O'Neill was a particle physicist at Princeton University, and was best known for his work on linear accelerators, better known as atom smashers. These devices accelerate charged particles to very high speeds in electromagnetic fields, and observe the debris resulting from their collisions. In 1969, in the wake of the Apollo 11 lunar mission, he asked his students to consider the following question "Is a planetary surface the best place for an expanding technological civilization?" The conclusion reached over the summer was, "No". For access to clean free energy, raw materials and room for growth, O'Neill and his class concluded that free space was far more suitable for an evolving civilization, and that, considering the recent proven practicability of space flight and accessibility of the Moon, together with the industrial assets of weightlessness and vacuum in the construction of very large objects, it would be possible to build large Earth-

like habitats using technologies either available at the time, albeit scaled up, or at least easily foreseeable. For 5 years, he struggled to get his ideas for space colonization into print, finally publishing an article in "Physics Today", in 1974, shortly to be followed by a book "The High Frontier". The results were amazing, in that there was a massive surge of interest from the general public in many nations. He realized early on that, because space settlement implied the participation of a much wider constituency than conventional space activities, this was an idea with a potential mass appeal. After a series of conferences with N.A.S.A in 1974-77, it became clear that, although interested, N.A.S.A simply could not take on the long-term research and development, and that several critical areas of research could not be funded. The Shuttle programme, and, later, the Space Station, was and is likely to require their full attention. Accordingly, O'Neill, and two associates at Princeton, the Yorkshire born physicist Professor Freeman Dyson, and Dr.Henry Kolm, set out to identify "critical path" research which would facilitate the move towards Island One and its associated Solar Power Satellite industry, and which, furthermore, could be demonstrated on the laboratory bench at low cost. In 1977, therefore, they founded the Space Studies Institute, which every two years ever since, has hosted conference on space manufacturing, at which the results of their work, and the development of ideas on the frontiers of space development, are widely discussed. This Institute is run by a non-elected board of trustees, which manages work on a series of focussed goals, all of which are designed to make feasible, or more economic, the eventual settlement of space. They have funded research into, and, in appropriate cases demonstrated models of, mass drivers, lunar materials extraction plants, and oxygen concentrators. They have produced ever lighter and cheaper designs of S.P.S, new space craft designs, ways of using Shuttle External tanks, and sat on numerous Government Commissions on space issues. Senior associates of the Space Studies Institute have been instrumental in founding the International Space University, which now offers degree courses in space sciences and technologies, and is already producing a future generation of leaders and thinkers heavily influenced by the ideas of O'Neill. These will increasingly find their way into industries, civil services and governments throughout the industrialized world. Expansion into space has been adopted as an official goal of America by the House of Representatives earlier in the 1990's.

Early Space Settlements

Island One, as conceived by O'Neill, is a sphere of about 400-500 metres in diameter, with a long central hollow shaft as an axis. At each end of the axis, external to the sphere, are docking ports, and microgravity workshops and industrial plant. Also external to the sphere, between it and the industrial ends of the axis, will be transparent plexiglass, doughnut shaped tyres 400-500 or so meters in diameter, linked to the axis by small hollow tubular spokes, which will give access from the hollow axis. These "tyres" would be the agricultural areas and would receive eternal sunshine in precisely regulated intensity and wavelength. The entire structure would be rotated about the long axle to create a centrifugal force on the sphere's surface and agricultural tyres. This force is similar to that obtained by whirling a bucket of water in a circle about one's head; if fast enough, the resulting force keeps the water in the bucket.

This would provide artificial gravity for the colonists, who live in the sphere, and the agricultural areas, if desired, while along the axis and at its ends, the gravity would be zero, so that industry would be carried on in microgravity. The artificial gravity need not be 1 G, as on Earth's surface; if so, an Island One of these dimensions would have to rotate at two revolutions per minute, which could cause giddiness. One cycle per minute is considered ample for health grounds, and slow enough for comfort. Accordingly, a gravity of 0.2 to 0.3 G would probably be considered sufficient, and would enable the average human to weigh an agreeable 20-30 kilos. This would probably suffice to prevent the worst hazards of re-adaptation; in any event, a low Earth orbit large space station would easily provide over 2-3 weeks, an increasing range of gravity from, say, 0.2 to 0.6 G, so that re-entry to Earth's conditions would be easy. For a smaller space station, this degree of gravity could be provided by attaching the station to a tether, 1-2 kilometres long, attached to a counter weight-e.g. another space station-and rotating the tethered pair about the centre of the tether, like a dumb-bell. Alternatively, the Low Earth Orbit station could consist of a series of concentric rings set in rotation about an axis, or, again, a cylinder whose axis points towards the centre of a notional wheel, of which the cylinder forms part of an imaginary spoke. Either system could provide artificial gravity to any desired level. The station need not be of enormous size, but could easily act as a "re-adaptation unit" if necessary.

Experiments by agricultural experts have already shown that in intensive hydroponic farming it is possible to feed over 50 persons per from an acre of land; with perpetual sunshine, no weather, complete elimination of pests and diseases, and skilful selection of crop regimes, Island One will be able to feed 100 people from an acre of hydroponic farm. Each agricultural "tyre" will have a surface area of at least 10 acres if cultivated along half its cross sectional circumference, assuming a tyre diameter of 400 metres, and a 10 metre depth. Thus 10 tyres stacked in parallel will feed the entire population of Island One. The initial food, plant seedlings, fertilizers, and air will be supplied from the Earth, Moon, and asteroids, but, eventually, there will be almost total recycling. The plants will also produce part of the colonists' oxygen requirements, as on Earth, along with lunar soil, and will use water recycled from the colonists and the industries. Most familiar and well loved crops have already been shown to grow well in microgravity conditions, and meat could eventually be provided from rabbits fed on alfalfa, chickens fed on scraps, or fish farming. The full cultivation of wheat has now been demonstrated on Mir by 1997, while Japanese researchers are working on self contained fish farming systems for their beloved sushi and sashimi dishes.

In 1997, NASA based experiments in closed-cycle prototype lunar bases had reached the stage where a 6 metre diameter cylindrical habitat supported 4 persons for 90 days with extensive recycling of water and oxygen, with considerable food production. Future work aims at raising self-sufficiency towards 100% for all three items (closed loop recycling) for indefinite duration, and then reducting size and complexity of the habitat for future use in Space.

Inside the sphere, the colonists will live and relax. The 5-10, 000 inhabitants will have access to a land area, on the inner surface of the sphere, equal to about 450,000 square metres, or, if one wishes too exclude the "polar" regions as having too little gravity, about 300,000 square metres-which will consist entirely of homes and gardens, as all the industry and working agriculture will be separated from the living space. Each person will thus have 30 square metres of living space, or, assuming that contract workers would come as a family unit of 2-4 people, that would mean about 100 square metres per family unit of ground space. This would compare very favourably with earthly conditions for many people, but on top of this there would be a considerable bonus. The third dimension would be available in a way unknown to most terrestrials. There would be a cylindrical volume of space 400-500 metres long with a diameter of maybe 100 metres in which gravity would be virtually non-existent.

Aerial sports could multiply; obvious examples are flying with unaided muscle power-simple one metre long nylon wings could be used effectively even by the unfit. This would fulfil the age-long human dream of care-free flight. Zero gravity sports, diving spherical swimming pools would be a whole new experience, as would ballet. It was observed in 1910 by an ardent admirer of the great Nijinsky that his landings were so graceful and delicate as to create an impression that he was actually retarding his fall. It is not hard to imagine whole new schools of ballet, gymnastics, and even circus acts in the unique environment of Island One.

Much ribald comment has appeared on the subject of zero-gravity honeymoon hotels-I merely suggest that the incumbents of Mr. Conrad Hilton's empire look carefully into the prospects! Another distinct possibility for Island One type settlements is put forward by Dr. Kraft Ehricke, and, as a doctor, I have a natural interest. We have seen earlier how low gravity exerts less strain on the musculoskeletal and cardiovascular systems; people affected by severe heart failure, or rheumatic and bony diseases might welcome the freedom offered by 0-0.3 G settlements. Many, aged or crippled by arthritis suffer precisely because of restricted mobility. This mobility restriction may arise either from sheer pain, or from the wasting of muscles, or rigidity of the joints. Similar problems afflict children with limb spasticity or spina bifida. Many of these have normally functioning intellects, and some even have superior ones. Microelectronics is coming increasingly to the aid of those who can think, but not communicate; coupled with reduced gravity, the use of aids to movement would surely be much simpler. Further, many cases of "incontinence" result purely from inability to reach the bathroom in time, rather than from unawareness of the need. A period at reduced gravity will be a boon; it may be that one or two months per year will provide such a boost in morale, and exercise of muscles long neglected due to the hardships of gravity, as to refresh them for the rest of the year. Let us think back a mere century when tuberculosis of the lungs was not uncommon in Europe, even among people from "gentle" backgrounds. Before anti-tuberculous therapy, many people were advised by their physicians to go to Switzerland for the clear pure rarified air of the Swiss Alps. Although this was not a cure, it did provide a rest from the polluted atmosphere of industrial cities, a change of scene, and, in some cases, allowed the body's immunology to contain the illness for long periods. Many people felt much better for their experiences. Thomas Mann's novel "The Magic Mountain" gives a poetic impression of the way these people lived. Such treatments were costly-months in a Swiss chalet would have been expensive for a family based in the North of England in the 1890's; yet it was an accepted recommendation for people with means.

If the history of air travel is any guide, then by 2020-2030, the cost of space travel will fall considerably, to a level comparable with first class trans-Atlantic flight today. Indeed, young people who start annuities of long-term savings schemes may well be able, by the time they reach their late sixties, to contemplate the fare to Island One, or to its successors.

Island One is most likely to be built by a consortium of many interests-utility companies, industrialists, governments of several nations, science foundations, pharmaceutical manufacturers, biological research institutes, tourist organizations, and so on; the inhabitants are likely to be of diverse social, cultural, and religious backgrounds. It is an open question whether these differences will lead to rivalries within the community, or whether the opposite will occur. The successful functioning of such a colony will only be possible if people do in fact share a common purpose, and a common ideal of service to humanity as opposed to some small and arbitrary part of it. An overdeveloped sense of national particularism is likely to be inefficient, and, in the interests of cost effectiveness will need to be weeded out in the selection process.

Experience in Russian long-stay missions suggest that the crew may well develop a sense of solidarity, perhaps even directed against their controllers on the ground. They will face the problems and opportunities for comradeship among people in a restricted and closed environment throughout the ages.

There are other trends in society today working towards the dilution of national boundaries and sentiments-the electronics and satellite revolutions are already enabling worldwide and instantaneous communication of data, ideas, and culture. Many musicians, writers, artists, scientists, and travellers who consider themselves as members of a fraternity that transcends mere national boundaries. As of 1980, there were over 3,900 transnational organizations or interest groupings, as compared with 1,000 in 1960. This process of "transnationalization" is likely to grow, and can only lead to a deflation of nationalism in the future. There will doubtless be outpourings of nationalism as embittered bigots try to resist the coils of the global market economy-Iran is a case in point-but their reward is likely to be a well deserved beggary. The Internet is likely to subvert still further the concept that a person be defined by his or her nationality. Considering the past 100 years, it is hard to view this process with anything less than unclouded enthusiasm. Among skilled and professional people employed on a common task of glamour, prestige, challenge, excitement, and value, I suspect that national and religious rivalries will come to be treated with indifference, and so I am hopeful that Island One will prove an ultimate solvent of "childhood" rivalries. Only talent, professionalism, and character will mark out people from each other in the stimulating new worlds of tomorrow.

Island One, then, will be a little like a village community of 5-10,000, but a village like no other in the history of the world. For it will be multinational and multi-talented. The value of their manufactured goods will exceed that of most individual nineteenth century nation states, and the colonists will be in constant communication with family and friends back on Earth, or on the Moon, each 400,000 kilometres away. They will thus be an integral contributor to the common culture of Humanity. The isolation of rural life will have no meaning. Their work, and their recreation, will be within walking distance, yet no noise or smoke of industry will disturb them. They will live in a culturally rich environment, with every conceivable electronic library, data base, or entertainment available at the flick of a switch. They will have unfettered, unrivalled opportunities to study and utilize their new medium, the Ocean of space, with its stars, planets, radiations, and other unknown features. Every wavelength will be open to them, and they will, in time, design and fit out probes and expeditions to the far reaches of the solar system and beyond.

Yes, Island One will be a village with its own community life, but, unlike so many others in the long history of human civilization, it will be isolated by distance, but not by time or communications problems.

The inhabitants will be, for the most part, intelligent, enterprising, courageous, and adaptable, in these early stages-why else would they be there? In effect, the Island One pioneers will be self-selected from the billions of Humanity, chosen by talent and enterprise. Those who go to Island One, and do not return in disillusion or beaten by life in a new environment, will gain an attachment to it and the way of life it makes possible. After setting up solar power satellites in new industries, some will return to tell people on Earth of their life up there. Inevitably, among thousands, probably millions of skilled and enterprising people on Earth, frustrated by lack of opportunity, recession or uncongenial and restrictive environments there will arise a curiosity, and a desire to see for themselves the New World of Island One. There will be books, television programmes, life stories, and, eventually, conducted tours showing life on the new frontier.

The works of Dr. Gerard O'Neill have already inspired thousands to work for Island One, even many who will not live to see it, or will be too old to reach it. How much more so, then, when the prospect of Island One becomes imminent, and the ecological crisis on Earth has deepened, with the inevitable collapse but a few years off? So it is that by the second quarter of the next century that the migration will commence. By this time the Moon-L5/Island One-solar power satellite economy will be in full swing, with Island One the lynch-pin of human civilization. Lunar industries will have expanded considerably, and there will be many mass drivers working in the S.P.S manufacturing programme. The advent of self-replicating 95% automatic factories will further enhance the capacity of space industry for truly gargantuan feats of engineering. Even in the 1990's, Japanese factories can turn raw steel into motor cars with minimal human presence. The energy-poor billions of Earth will be clamouring for ever more satellites, and Island One will be earning trillions of dollars.

A New Home for Life

Dr. O'Neill called the next stage "The Humanization of Space". Starting with Island One, he saw an irreversible manufacturing base in space, growing exponentially, since economic growth in manufacturing can arise, from a sound base, far more rapidly than on Earth. There are no energy restrictions, no shortage of raw materials, and no environmental pollution. Large constructions will be built semi-automatically in very short times owing to the lack of gravitational or frictional forces. The only constraint on the humanization of space will be, at first, the availability of human beings! So, by about 2050, he envisaged the next stage of settlement-a true space colony, of maybe 100,000 to 200,000 inhabitants. This colony-Island Two-would consist of a pair of parallel cylinders each 6 kilometres long with a diameter of 1.5 kilometres. Each of the cylinders would rotate about its axis, and transport between them would be provided by small vehicles released from the rotating surface at predetermined moments so as to reach its neighbouring cylinder. Each cylinder would have 3 longitudinal windows, and would be illuminated by 3 reflecting panels attached to the end of the cylinder, set at an angle to the axis. The intensity of the sunlight in the cylinders would be set by the angle of the reflector panels, which could be opened and closed like the petals of a flower; the panels would be the flower, and the cylinder the stamen. These cylinders would rotate, so that the entire interior surface would be subject to centrifugal force, while the axial region would be a microgravity region. Once again, industry and agriculture would be arranged in separate facilities outside the residential cylinders. The habitable surface area would be about 30 square kilometres in each cylinder; for each of 100,000 persons this would provide about 300 square metres, or, in the case of a family of four, about 1200 square metres.

On these figures, with the agricultural and industrial ones outside, there will be room for large open areas of common land, to be planted, perhaps, as woods, parklands, lakes or hills. The choice of terrain would be largely optional, as indeed would be the climate. Length of day, brightness, temperature, and scenery would all be matters of taste. It would, for instance, be totally feasible with the level of the L5 economy to build a series of Island Two colonies, of differing characteristics. New Hawaii, New Switzerland, new San Antonio, and New Corfu could exist in nearly perfect replicas in each of several Islands in space. If this prospect becomes realizable, millions, like their forefathers before them, would seek new lives and opportunities in the celestial New Worlds. The Colonies will not arise unwanted, nor be filled by some terrible process of conscription or penological coercion. If there is any selection process involved, it will be self-selection by able and willing volunteers from all the peoples of the Earth. I am sometimes asked, conversely, how, in a deteriorating terrestrial situation, people will be selected to go to the colonies, rather than how will unwilling masses be compelled to go. If popular demand exceeds supply, is there anything a person could do now to enhance their own, or their descendants', chances of being allowed to settle in a space colony. To such an enquirer, I would reply that, apart from the obvious desirability of acquiring a trade or skill likely to be of value in a community, a proven commitment to the idea of building a human(e) future in space will be a trump card. Just as descendants of the Pilgrim Fathers constitute a sort of aristocracy in the United States, members, or descendants of members, of current pro-space organizations, will be likely to enjoy high status in the settlements, the ultimate fruit of their labours. Those people, therefore, who fear for the future of terrestrial civilization, could well do a great deal for their children in actively supporting the humanization of space. They are also likely to acquire the sort of long-term fame due to pioneers at all times. There could be real personal benefits in supporting, actively and publicly, the human expansion into space; whereas, if this development does not happen, there will, I fear, no future worthy of aspiration. Either way, interested readers can lose nothing by giving support.

No, the problem will be to provide the means of mass transportation, and the Islands themselves, in sufficient profusion to cater for the demand. For, although Island One will be a hive of highly technical and specialized industries requiring skilled scientists, engineers, construction workers, biomedical specialists and systems analysts, the Island Two stage will call for all the trades and skills from the sophisticated research scientists to the artisans, entertainers, and tradesmen who make up the fabric of a rich and complex civilization. Firstly, rapid mass transit systems from Earth to orbit will be required. This will be the most difficult part of the whole humanization of space, and will require considerable advances in SSTO's. When I first became aware of this subject, this aspect seemed a real poser, but, prospects for a solution during the second quarter of the coming century are much brighter than even 5 years ago. From the DC-3 Dakota to the DC-10 took less time than I envisage for the Island Two stage in space. It could even be that an operational X-33/Venturestar, or a similar re-usable vehicle derived from X vehicle technology (c. 2010?) will be the last spaceship built on Earth, and that its successors will actually be made in space factories.

The current plan for Venturestar could perhaps carry 50 people to Earth orbit, maybe 100 times per year. A fleet of 10 could thus take more than the population of Island One into orbit each year by 2020-30. An Orbital Transfer Vehicle, using lunar derived oxygen and powdered aluminium, could well be available at the same time. A fleet of 20 could make the weekly round trip to L5, thus again populating Island One in a year or so. Given that we are looking to nearer 2030/40 for Island One's inital construction, this does not seem an excessive traffic model for the first generation of operational third generation spaceships and derivative O.T.Vs. The second generation vehicles, carrying perhaps 100-200, like our present medium jets, could easily, on the model of civilian aviation, arrive in time for Island Two.

More futuristically, the Russian scientist, Artsutanov, described an idea which could provide the ultimate low cost mass transit to space. He proposed a gigantic tower or lift which would be lowered from a high orbital platform to the ground like a spider's web, and act like a giant lift shaft. Such a construction would need a material much stronger and more resilient than the hardest steel, and even the revolutionary material Kevlar falls short-by 5 fold-but, a century from now- who knows? Recently, a variation of carbon based "bucky ball" chemistry yielded 12 times the strength of steel for much less weight, while nanotechnologists believe that assembling structures atom by atom, creating in effect a single shaped diamond of any size or shape, could be attainable within 30-40 years, probably in space! These would allow vast improvements in weight and durability of third/fourth generation space ships. As for Orbital Transfer Vehicles, O'Neill believed that larger ships, could carry hundreds if not thousands of colonists by electric-ion drive, over several. A mass driver engine, using compacted pellets of cosmic slag, propelled by solar electricity, could equal a high-energy upper stage in efficiency, and derive its fuel from the waste products of space-based mining. This beautiful concept is a very simple, and has virtually no running costs at all. With an 1800G exhaust velocity already demonstrated, it could provide an ideal reasonably rapid bulk transport system which could, in time, be the key to the entire solar system. Mass driver powered vehicles would ply the vastness between the worlds without ever landing, and would collect their mass fuel from any of the innumerable minor rocks and asteroids that litter interplanetary space. Thus, on a mission to the outer planets, fuel could be mined and picked up on site. In many cases, this material would also be payload and resources for use in industry. Each small asteroid or pile of rubble mined or compacted into mass driver pellets represents the elimination of a potential Earth collision disaster of the kind described earlier in this book. It is not often that one can kill three birds with one stone at no extra charge! Initially these vehicles would ply the routes from Earth orbit to L5 and lunar orbit, but eventually ships would range from Phobos and Deimos, the moons of Mars, and the asteroid belt. Delta-V from L5 to Ceres, the largest asteroid, would be very low- only 2-3 Km./second, as compared with 11.2 from the Earth's surface. This brings us to another fact about the L5 colony. There would be two different but not mutually exclusive directions after the attainment of Island One based industry. Firstly, there is the humanization of space, by way of larger colonies such as Island Two, and the larger Island Three; this last is the full-scale O'Neill Colony, 32 kilometres long with a 6 kilometres diameter with several million inhabitants. This type of colony could appear in the second half of the 21st century, and would begin to make inroads into the Earth's population at this period. The second way forward is to the planets. With flourishing Island One type colonies it will not prove difficult to launch major expeditions to the planets with little or no involvement by the tax-payers of Earth. For instance, base camp could be set up on the Martian moons using Island One techniques; from this base landing operations could be established on Mars using locally built vehicles and fuel depots. If desired, the climate of Mars could be modified by terraforming methods described by James Oberg in his book "Mission to Mars". The genetic engineering techniques available on Island One-rapid mutation by exposure to cosmic radiation, gene splicing and purification methods-could be used to produce super-resilient blue-green algae to alter the atmosphere of Venus by the scheme outlined by Carl Sagan. Current water shortages could be alleviated by imports from small comets. In time, Venus could be rendered habitable?

Crewed expeditions, to the asteroids, outer Galilean, Saturnian, and Uranian satellite systems could be fitted out, with appropriate radiation shielding, from the L5 industrial base at Island One. The Voyager missions have shown that these regions are far more diverse and interesting than had previously been imagined, and an understanding of these "mini solar systems" will be of vital importance in learning how solar systems as a whole are formed and how they function. Jupiter's moon Europa is of especial interest; as described by Arthur C. Clarke in "2010: Odyssey Two", it is literally an oasis in the desert, being covered by an ocean of water 50-100 kilometres deep, and overlaid with a thin layer of water ice. The presence of liquid water so far from the Sun is believed to be possible because of the tidal heating of Europa's core by the massive gravitational pull of Jupiter, only 600,000 kilometres away. Since Europa has about the same mass as our Moon, with an escape velocity to match, it is not hard to see in Europa a fuelling and supply base for the entire outer solar system- provided that Europa is not already inhabited. Some scientists have even speculated about marine life in the warm ocean depths of Europa. The other moons, Callisto and Ganymede are also covered with water ice, as are the Saturnian moons Rhea, Mimas, and Enceladus. The long trips to the outer solar system may be compensated by free water, oxygen, and rocket fuel on arrival. This completely alters the economics of the solar system for human occupation and development. Island One style colonies, designed for a few dozen crew, could set off on long voyages of discovery like Cook and Magellan, taking several years to explore the far reaches of our solar system. Scientific or mining bases could well be set up on many of the outer planet satellites, asteroids, and comets during the later years of the next century; indeed, barring a major collapse of civilization or failure on the part of humanity to build up space industries and Island One, this migration phase seems highly probable, and consistent with human nature and history.

Free Space or Planets?

There are space scientists and enthusiasts who place greater emphasis on planetary colonization and development, and others who favour the massive build up of L5 type colonies as an ultimate solution to the problems on Earth. They cite the fact that Earth's population is bound to keep increasing for at last 50 years before stabilizing at a minimal level of 10 billions by AD 2050. The lesson of history is that zero population growth does not occur before certain standards of living are attained, unless there is to be a great increase in the death rate. To achieve zero population growth rate we must raise the living standards of the less developed countries with all the problems discussed earlier.

My own view is that the two programmes are interlocking, but that the Moon-Island One/L5-Solar Power Satellite economy is the essential first step; for only by this route can we tap the resources of space quickly and economically enough for the peoples of Earth. It is really only thus that we shall reach the stage of having reasonable choices in the face of the population/energy crisis looming over the horizon of the 21st century. The sooner this scenario is implemented, the more people will be able to benefit from the new opportunities offered by Island One. To ensure that a significant part of Earth's population has the choice of migration to the

New Worlds, it is really important that space development is pushed forward, with public support, with some urgency. Leaving it to the next generation on grounds of cost or whatever will merely condemn many of today's children to a world of diminishing opportunity, and deprivation.

Furthermore , we cannot know that our skills and engineering base will not atrophy if not kept in use, so that should an unexpected natural disaster threaten us-like a burst of supervolcanism or a new long period comet impact-it will not find our civilization confined on one small planet. What we have in space capablity we use and develop, or lose. Life is a race in which he who does not will to win, deserves to lose.

Island One, therefore, is the jumping-off point for the future-the humanization of space and the settlement of the solar system. Without it, none of this will happen, and human civilization will falter and decline, in the lifetime of today's children, into a Dark Age of unknowable duration.

The colonies of Island Two and Three will allow, initially, an enormous mixture of talents, race, backgrounds, nationalities and creeds, all drawn together by the lure of a new life on a vast frontier. Later, however, I believe, there will be a re-creation of diversity in human affairs, each colony will become more or less autonomous, with its own chosen climate, "geography", political and religious constitution, and even in length of day and night.

One colony might resemble a Greek island, while other might resemble Austrian mountain communities, or small English market towns. There is really no limit save the collective imagination of the colonists. The larger colonies, with their artificial gravities, can have winding streams and lakes stocked with fish. All manner of sports and hobbies could flourish, and even wildlife threatened on Earth could be conserved. Similarly, many unwanted species of Earth, like tse-tse flies, mosquitoes, blowflies, and soccer hooligans could be excluded ab initio, never again to trouble the children of Man.

What would people do in such colonies? Why, most of the things they do on Earth, but in better surroundings. They would work in the new industries, or the services which will spring up around them. There will be new farms, crop development, leisure industries, and universities of excellence. The sciences of astronomy, planetary sciences, computerized transportation and many others will flourish. The new art forms of large scale landscaping, climate design, planetary engineering, and total ecological management would develop, and tax the creative ingenuity of our children. Biomedical sciences would make a quantum leap forward-infectious diseases could perhaps be stamped out at source. Many genetic conditions will be increasingly identified at conception and corrected by the new genetic surgery. It seems likely that cardiac and rheumatic diseases will be greatly reduced, and, according to some biological specialists in the E.S.A, healthy human life-span could eventually double to 150 years. The supporting occupations-the performing arts, entertainments, culinary, sports and so on would accompany the frontierspeople as in the course of human history. 100 years from now, perhaps the strains of Gilbert and Sullivan's operas will amuse the puzzled inhabitants of a distant Island in the Sky.

America was colonized quite some time after discovery; after the frontierspeople, entrepreneurs, and heroes came the artists, barbers, chefs, and poets. No-one asked in 1700-1800-"what will they all do?" They did what people have always done-their own thing. In the colonies at L4 and L5 (Lagrangia) there will be no want of food, no troublesome weather, and very little dull repetitive work; but there will be arts and crafts, hobbies and sports, community life, and enough mental challenges for the most highly developed intellect. By a century from now, Dr.O'Neill believed that a substantial proportion of humanity will be living rich, fulfilling lives in space colonies of every conceivable size, construction, landscape and constitution; and yet they will all be human beings, linked by the marvels of electronics and the laws of celestial mechanics. Also, by this time, it will be possible for communities of different religious or political persuasions to seek a new life without interference to or from the rest of Humanity.

Lest people consider this farfetched, let them recall how many settlements in America arose to avoid political or religious persecution.

What pressures will drive millions of Earth people to choose the beckoning vastness of Lagrangia and beyond? I believe the same drivers as have operated throughout history; firstly, those vigorous and constructively rebellious young who in all times have turned their backs on overcrowded cities or conformist societies to make their own way; secondly, the explorers driven by curiosity and the search for new worlds; thirdly, the entrepreneurs who move with boldness and resolution to open up a new market, and finally millions who, each for their own reasons, have decided that the future has no more to offer them at home, or who feel that they can do better for their loved ones elsewhere. Who is to say that religious and national persecutions have ended, and that the 21st century will not see them again? Certainly not the Bahá'ís of Iran, who have experienced the full cruelties of fanatical intolerance. As their religion becomes more widely known, and is more generally perceived to be opposed to most of the world's vanities and delusions-notably nationalism, sectarianism, racialism and sexual chauvinism-and faces persecution from entrenched bigotry, many Bahá'ís may find themselves drawn to the idea of migration to new worlds. They will be peculiarly suited to work constructively to bring about the new civilization which Lagrangia promises. The elevation of talent and educational attainments above all considerations of national or subgroup loyalty, their reverence for the boundless works of God as seen in His cosmos, and the necessity for tolerance and fairmindedness in resolving internal disputes, would all provide good criteria for the selection of colonists in space!

Unity in Diversity

The new colonies and the subsequent humanization of space could provide the perfect material substrate for a Bahá'í civilization, with Unity in Diversity as its watchword. The colonies, although diverse in lifestyles and culture, would recognize the underlying unity of all men and women under the ever-watchful intelligence of God.

The colonies, with their relatively small populations, mostly known to each other, will be ideally placed to fulfil another age-old dream; that of true participatory democracy. Such a democracy has only ever existed for about one generation-namely the Age of Pericles in Athens during the last third of the 5th century B.C. In this city the 10,000 free citizens met in the market place to discuss and debate all political decisions of peace or war or public expenditure taken by the city, while the rulers were all elected annually by the entire free citizenry; non-participating citizens who led private non-political lives were called "Idiotes"-hence our word. However, the bulk of the 400,000 strong population had no voice at all-they were slaves without rights of any sort.

In the space colonies, the situation would be truly remarkable. There would be a relatively small, educated or skilled population with access to information and resources from the entire solar system. The colonists would build a true community, whose members would know nearly their entire population, at least by sight or greeting, and the only slaves would be mechanical-electrical, without human rights to be violated. The population would have in common a desire to build and operate a new community in space; indeed any migrants who discover that they have made a mistake, will be perfectly free to return to Earth, or even move to a different colony. For the colonies will not be isolated, and there will be flourishing trade and tourist routes throughout civilization. A true Athenian style democracy could thus flourish again in the distant recesses of space. Or, then again, there could exist a colony of Platonic philosophers ruled by a Philosopher-King. Both could co-exist without detriment to each other. If desired, their separation in space could be so great that neither could ever be a military threat to the other.

It would be rash to claim that warfare would be ruled out, or that human nature would suddenly give up ideas of war and domination; however, there are some encouraging points to make. Firstly, the humanization of space can tap the energy and resources of the entire solar system. Dr. O'Neill has shown that this could support a human population of 3 million times the present Earthly population at a level of prosperity equal to that of a Hollywood film star, if so desired! Even with no birth control, this gives us plenty of room for manoeuvre-at least 12,000 years; and probably nearer to a million if zero population growth is attained. High living standards will help to achieve this. The competition for resources and room is certainly one major cause for warfare, and this would no longer be operative.

Secondly, distances between colonies would make warfare very difficult logistically. Assuming that a colony wished to make total war on another, it would wish to be able to threaten total annihilation. Any type of nuclear missile would be worthless, as it could easily be detected thousands of kilometres before arrival, and peremptorily destroyed by high energy laser beams. A high energy laser beam itself would take a finite time to reach its target, and would only be able to burn a small hole through the wall of the structure. A small hole would not result in a catastrophe-the atmosphere would take days or weeks to leak out, and would be detected and plugged within an hour. A light-reflecting surface could then be interposed in the direction of the known enemy.

Political and religious differences could be motives for warfare as on Earth, but it is hoped that the dispersal of Humanity into groups of nomadic mutually trading colonies throughout the solar system might diminish the desire of one habitat to impose its beliefs on another. Distance should dilute bigotry. In the longer run, the fact that most of the settlers will be skilled and educated people, and that these attributes will be highly prized and nurtured among future generations, should create a marked bias against bigotry and fanaticism. If one group of people differ in lifestyle from another group 10 million kilometres away, it is unlikely to create more than a polite interest.

The dispersal of humanity into hundreds of different colonies does at least multiply the chances of survival and development of human civilization; the prospects are much better than on Earth, with the shadow of nuclear winter or asteroid impact induced climatic catastrophe hanging over every man, woman and child clinging precariously to one small blue planet. Space settlements will not solve all human problems, but will buy us time and opportunities to make the attempt. Without them, our near descendants are very unlikely to solve the problems in the short time available-we really have nothing to lose by going for space in a major way. If my prognostications for an Earthbound future are wrong, we will have acted unnecessarily, but won a Universe. If I am right, we will have saved human civilization, if not the species itself, from Earth-bound decline and extinction, and added greatly to the self-creative potential of an Embryonic Universe. Diaspora is a proven strategy, both in history and evolution, for survival and onward development

In summary, then, the Humanization of space is a feasible and logical outcome of the Island One phase of space industrialization; it is could be under way within half a century, and lead humanity away from the brink of collapse to a remarkable civilization, more unified yet more diverse, and more truly humane than any that have graced the history books of Humankind.

Chapter 10

The Universalization of Humanity.

We have now reached the remarkable conclusion that, if we make the right choices, avoiding the hazards of nuclear winter, and the ruin of civilization through over-population and ecological collapse, our now living children, together with many adults alive today, could be witness to the Humanization of space by the middle of the incoming century, and will be able to see the course of human history unfolding. How far could the Humanization of space go?

The Russian astronomer, Nikolai Kardashev, describes three stages in the growth of a civilization. In Stage One, which obtained up to the year 1957, the young civilization lives off the resources of the one home planet, the planet of its biological birth. We terrestrials are at the dawn of Stage Two, in which the emerging civilization utilizes the resources of its entire solar system-or, put another way, taps the resources of its parent star. The third, or mature, Stage Three is a Universal stage, in which interstellar travel and migration have come, and the civilization moves out to populate its own galaxy.

The L4 and L5 regions occupy 24,000 billion cubic kilometres of space. If each Island One is 1,000 kms away from its neighbour, there can be up to 600 million people in Lagrangia. Each habitat would be as visible from Earth as a 2 pence piece is at a distance of 5 kilometres, and from each other, would appear like a 2 pence piece at 30 metres. Later on, between the orbits of Venus and Jupiter, there could be an inner belt, 1 million kilometres thick, covering about 625 billion sq.kilometres, or 625 quintillion cubic kilometres. If each Island is to be 1,000,000 kms from its neighbour, this allows 1,300,000 colonies. If these are Island Two types, this allows the settlement of 300 billion people in between the orbits of Venus and Jupiter-50 times our present numbers, with each 200,000 strong settlement occupying a sphere of space 1 million kilometres in diameter ! These colonies would be invisible to Earth, or each other, with the naked eye-like a 2 pence piece seen at 4 kilometres distance.

The solar collecting surface must increase in radius directly as distance increases from the Sun: thus, since the Sun's energy available for collection at Earth's orbital distance is 1.3 kilowatt per square metre, a 1 sq. kilometre collector , collects 1.3 gigawatts of energy, roughly equivalent to 1.5 nuclear power stations-easily enough for Island One with 10,000 inhabitants. For Island Two settlements, with 200,000 inhabitants or so, in Earth's orbit, a 6 kilometre radius collector would deliver1.3 terawatts (1×10^{12} watts) of power-night and day. This is some 25 times the total electricity generation of the United Kingdom! Because the energy density of solar radiation falls off by an inverse square law, at double the distance from the Sun, the collecting area must be quadrupled, or the radius doubled. Thus, near Jupiter, the Island Two colonists need a collector of 24 kilometres radius to take the same energy as near-Earth residents. Beyond Pluto's dark regions, a collector some 300-400 kilometres in diameter would give full Earthlight and warmth to a colony, and so on. Island Two colonies would need about 50-60 million tons of materials, including shielding: the implied 1,300,000 colonies would thus require some 800,000 billion tons of material-a mere fraction of the matter available from the Moon and asteroid belt. Solar energy will flow at the rate quoted for 5-6 billion years or so.

The following figures give an indication of the wealth of raw materials involved in building a Kardashev Two stage civilization

Earth Mass $6*10^{21}$ tons

Moon Mass $7*10^{19}$ tons

Total Asteroids $6*10^{18}$ tons at least.

1.5 million Island Two colonies 1*10^{14} tons

In addition, there are over 30 minor planetary satellites, and untold inert comets. The Kuiper belt, at tens of billion kilometres from the Sun, is thought to contain billions of comets and minor planets, while the Öort cloud numbers trillions. There is more than enough wealth of material and energy in "empty space" to nurture an extraterrestrial society at any imaginable standard of living for many thousands of years.

During the coming millennium, our descendants will become masters of the solar system, with settlements in every region, out to and including those now shadowy outposts, Neptune and Pluto; for, by simply expanding the area of the solar collector panels, Island One type colonies can derive solar power anywhere within 4 light days (about 80,000 million kilometres) of the Sun, since the radiation diminishes according to the square of the distance. The "Limits to Growth" and its entire restrictive view of the world becomes redundant. In the light of known human nature, it was at best naive to imagine that Humanity would avoid ecological disaster by suddenly imposing upon itself limits to growth, or that they would willingly endure the draconian loss of liberty which such limits would entail; the good news for our younger generation is that these proposed limits are unnecessary, and that those zealots who would impose them can legitimately be resisted.

Humanity will be moving out into a rich, prosperous and very diverse ecological medium, in which we will accomplish almost anything to which we set our minds. This is in such stark contrast to the Earth-bound alternative as to be scarcely imaginable to generations growing up in the 1970's and 1980's, who have been taught to see limits to growth everywhere, with only diminishing prospects in sight. Having begun on the Humanization of space, and arrived at a post-ecological crisis stage, we can now speculate a little more widely, and consider where the Humanization of space may bring us to. For we must now view Humanity's future, not in the decades, but in countless millennia, for a dispersed civilization, runs very little risk of sudden collapse through war or ecological crisis to which we are used.

In the last 10-15 years, the idea of interstellar travel has begun to approach respectability; there are already 3,000 or more serious scientific papers on ways and means of achieving this. The closest to achieving some sort of feasibility is the use of the thermonuclear explosion, and there is already a detailed design of a vehicle which could fly-by our nearest stellar neighbours in a journey lasting 40-50 years. The British Interplanetary Society published in 1978 a study called Project Daedalus, which was a proposed uncrewed probe to Barnard's Star, 6 light years away. It would be in two stages, and eventually 500 tons of instruments and local probes would enter that system. Initially, it would weigh 45,000 tons, of which the bulk would be Helium 3, obtained from the atmosphere of Jupiter. This Helium 3 would be converted into solid 4 cm. diameter pellets, and caused to fuse by high-powered lasers, like a miniature hydrogen bomb. It would "fire" hundreds of thousands of these pellets in succession, eventually obtaining 12% of the speed of light (i.e. about 120 million kilometres per hour). It would reach Barnard's Star in about 45 years and could well be built within the new century. This Project Daedalus is about as detailed and realistic to us as the B.I.S design for a 3 stage lunar rocket seemed in 1939. We know that they were only 30 years too soon, and that they were surprisingly accurate. If Daedalus is not realized, it will only be because a better way will emerge.

Rod Hyde of the Lawrence Livermore University in California has already advanced studies of laser initiated fusion of deuterium pellets in the search for thermonuclear fusion reactors, and believes that a "star drive" capable of 15% of the speed of light could well result in time. Since 1978 the Project Daedalus scheme is already beginning to look obsolescent-not because interstellar travel looks far-fetched-but because of sensational advances in anti-matter research. In 1978 only single atoms of anti-matter, in the form of anti-hydrogen nuclei, had ever been produced, and they had lasted for vanishingly small periods of time. An ordinary atom of hydrogen contains a massive positively charged single nuclear particle (a proton), with a very much smaller (1/2000th) negatively charged particle (an electron) in orbit around the nucleus. It is these electrons which determine all of chemistry and electromagnetism. In the anti-hydrogen atom, the heavy proton carries a negative charge (anti-proton), while the electron becomes a light, positively charged positron. When matter meets anti-matter, there is total annihilation, and the entire combined mass is converted into radiant energy. This is 200 times as much energy as is released in the fusion of two hydrogen nuclei into helium, since only 0.5% of mass is destroyed in this reaction. Anti-matter is therefore the most powerful store of energy known, or, at present, even conceivable.

During the period 1978-83 particle physicists began to use more energetic reactions in their search for the unifying factor underlying the weak nuclear and electromagnetic forces-two of the four fundamental forces of nature. This search for a Unified Field Theory, uniting the four forces of physics, has had some successes but in C.E.R.N and Fermilab, appreciable numbers of anti-protons were produced as a side reaction in the production of high energy protons for bombardment. The game is to hurl protons at a target, smashing up atoms and protons in the process, and to derive new insights into the structure of matter by analyzing the resulting debris. In 1978, physicists realized that, in producing anti-matter, they could deploy far greater energies than ever before simply by hurling beams of matter and anti-matter at each other and "picking up the pieces". They began, therefore, to syphon off and store these anti-protons for use in experiments. How could this be done, since, on contact with matter, there is a titanic explosion? The answer is that, since anti-protons are electrically charged, they can be moved and held in electromagnetic storage rings. Gigantic new super-cooled magnets were used and by 1983 sizable numbers of anti-protons have been marshalled into storage rings and held there for days at a time. They have even been linked to anti-electrons to make anti-hydrogen. Thus, over 5 years, relatively large amounts of anti-protons have been produced, cooled and stored at will. Amounts of up to one billionth of a gram have been handled as of 1983. Upon reaction with one billionth of a gram of ordinary matter, the resulting bang is as loud as a cap pistol. Thus, anti-matter is about 10-100 million times more powerful than gunpowder. The production of anti-matter in these experiments has only been a side reaction. If they tried in earnest, yields could be optimized for anti-matter. More recently, in 1994, another giant step was taken-anti-matter protons were removed and transported from their production site in a portable container; this represents a step towards more general applications for anti-matter-in particular, space transportation. As mentioned earlier, aluminium was a similar curiosity upon its discovery in 1826. I would think that the time interval between the discovery and utilization of aluminium would be comparable with the present stage of anti-matter development, and the relativistic interstellar spaceship driven by anti-matter. Today's younger Star Trek fans could well live to see the first interstellar space mission. If you reacted 180 kilos of anti-matter with 20 tons of hydrogen, and used the energy of annihilation to heat up the remaining 19.82 tons of hydrogen and derive thrust from it, a 5 ton payload could be sent to another star at over 25% of the speed of light. According to Robert Forward, of Hughes Aircraft Corporation, the percentage of light speed depends on the proportion of anti-matter to matter in the first stage of an interstellar rocket provided the payload to starting mass ratio is 1 to 5. The 5 ton payload in our above example could therefore contain 4 tons of hydrogen, with 30 kilos of anti-matter, which would finally place 1 ton of payload into the star system at normal interplanetary speed, after a rapid transit at a significant proportion of the speed of light. The trip to Barnard's Star by Forward's formula would therefore fall to 25 years, with an orbiter, not fly-by, at the end of the trip-unlike the 45,000 tons Daedalus which flies past.

Even in our present state of ignorance, we have come to the threshold of an imaginable method of interstellar travel. A mature solar system-wide civilization, with all the wealth of a thousand worlds at its command, can, over the course of the next few centuries, solve the still formidable engineering and technical problems involved in realizing Forward's ideas. Anti-matter, for instance, could be produced in hundred kilogramme amounts by utilizing the stream of protons which pour out in the solar wind at speeds of up to 10 million kilometres per hour. Perhaps, in an orbit near Mercury's, a huge collector of several hundred square kilometres could be set up as target to collect these protons, and the few anti-protons from these collisions could be saved and stored. Far-fetched, perhaps, but not beyond the resources of a Kardashev Stage Two civilization, evolving towards a Galactic community. This technology, married to the Island One Habitat economy, could make crewed exploration of the nearer stars with flight times of 10-20 years realizable. If this method is not in use within 100 years or so, it will be because a better way awaits discovery.

Another slower and more "natural" approach also suggests itself; since the energies required for interstellar travel are so enormous, why not use stellar energy itself? The Earth captures only 1/2 billionth of the energy emitted by the Sun. A sufficiently large mirror could catch and concentrate large amounts of solar energy, as well as using the minute pressure of light itself to accelerate a payload. This year, Sandia Laboratories announced a process whereby very thin films of a material related to polyvinyl chloride could be deployed in space from a small package and figured into any desired curvature by a computer controlled electron gun. An initial application envisaged for this technology-in about 15 years or so-is an extremely large space telescope, since the material composing the mirror would weigh 250 times less per unit area than the Hubble Space Telescope mirror. Over a few more decades, much might be achieved.

An O'Neill type colony, fitted with an onboard nuclear power plant for internal heating and energy needs, could afford to spend generations travelling across the insterstellar wastes, so that a propulsion system capable of achieving 1-5 % of the speed of light would suffice. The process would be akin to the ancient Polynesian island hopping, particularly since beyond

Neptune and Pluto there are countless thousands of small but materials-rich bodies in the Kuiper Belt, while the Oort Cloud, rich in comets, extends out half way to the nearest star. Our migrating descendants therefore will have stop-over points en route. Since O'Neill's Island colonies derive their sustenance from stellar radiant energy, and raw materials from asteroids and small moons, it does not matter if a trip to a star does not find a habitable planet, since the industries of the starship will be quite able to set up a local Kardashev Stage Two civilization from *any* local resources. Within a few years, the visiting colony would be able to create a local "Fermilab", or energy collecting mirror, build another O'Neill colony, and move on to another star, or make a return journey to our system. Thus, every 50-100 years or so, feedback from interstellar colonies would enrich the culture of the parent civilization-radio contact, of course, would be more frequent-perhaps every 10-20 years. Over a period of maybe 10 million years , the Galaxy may become woven into a kind of organic whole, with slower, more majestic rhythms than the frenetic pace of life seen in our times.

After Homo Sapiens...?

I have considered Humanity's likely impact on the worlds of our solar system, and, in due course, other solar systems, but we must not forget Newton's Third Law. In a larger, cosmic, sense it is possible that there will unfold an analogy here-in venturing thus into new and unfamiliar realms, will Humanity itself ultimately undergo changes? Will "Every action produces and equal and opposite reaction" apply to Humanity itself?

In building a civilization with access to vast resources of materials and energy, as well as giving societies enormous opportunities for cultural development without friction, we could well be greatly reducing the causes of strife and warfare among human beings. The new view of the Earth and the planets which the coming age of the Humanization of space will enlarge our view of the world and our place in the Universe just as surely as did the discoveries of Newton, Copernicus, Hubble, and Einstein , not forgetting the physical expansion of our viewpoints by Columbus and Magellan. The view of Humanity as a vast and diverse brotherhood, or, as the Bahá'ís have it, the Kingdom of Bahá, will receive a great stimulus from the view point of a space-faring culture.

But what of the innate drive to aggression and warfare which have characterized the recorded history of humanity? These have arisen and flourished because they have been rewarded by success. The emergence of Man as a dominant species owes a lot to the aggressive, pioneering, and restless spirit ; evolution has favoured these characteristics as the key to success. However, in the coming centuries, it is clear that these characteristics in their pristine form are no longer compatible with civilization. The virtues admired in an Alexander or Sennacherib are not to be admired in an Adolf Hitler or Saddam Hussein. These dangerous primitive emotions are not about to disappear in the short term, but there is the possibility that they will be sublimated. For the characteristic feature of the solar system is that, although there are enough worlds and challenges to satisfy the most expansionary and restless natures, this can all be done without the shedding of a single drop of blood. There will be no Montezuma on Mars, no Huayna Capac on the frozen wastes of Callisto and Enceladus. In time, with dilution and dispersal, the danger posed by illiterate hordes of hungry billions led by charismatic criminal lunatics will be a thing of the past.

In the longer term, evolution will work upon the delicate germ plasm of Humankind. For Man is and will remain a creature of biology, and, has not yet reached an end point of development. Most creeds have implied that Man can be perfected, either through Divine Grace, by biological evolution, or even by genetic engineering, to a more mature, adult creature in which his head, and his finer emotions, exert a more powerful and binding control over his lesser appetites. In other words, his nature is what we would call good not because of constant effort against backsliding, but because it is natural. Teilhard de Chardin calls this the Omega Man, a true partner of the beneficent Creator, and sees this clearly as a next step in biological and psycho-social evolution. This evolution is likely to take thousands of years, and will not take place in a Humanity restricted to Earth-there simply is not the time available.

In space, however, biological changes will take place in zero gravity. In essence, people will be taller, lighter, more graceful, and freed from most of the infections, environmental, and eventually genetic disease that now afflict us. Many degenerative processes will be greatly slowed down because of the 70% reduction in stress on the cardiovascular and musculoskeletal systems-some officials in E.L.G.R.A (the European Low Gravity Research Association) have already suggested on present evidence that our average life-span might exceed 150 years, and may well be better than this. This will make a 20-30 year journey in an Island One to the nearer stars much more acceptable than at first sight.

Evolution also teaches us that, when a new niche suddenly becomes available, as in, for example, the case of mammals after the extinction of the dinosaurs, or the first invasions of the land, there is a vast increase in the rate of development of new species to fill the niche left vacant. There is a great flurry of experimentation and diversification-Nature abhors a vacuum. In terms of history, J.D. Unwin has written that all great explosions in human thought and culture have been results of previous enlargements of his territorial or economic base. In this light, then, there is likely to be great and rapid diversification in human societies and, eventually, biology and psychology, when exposed to the challenges of a vast new and rich cultural medium.

Arthur Koestler has ascribed the human condition to the rapid development of some functions of the cerebral cortex which have not acquired the necessary neural connections to control the lower human brain functions and impulses. Perhaps evolution in space may provide a natural solution to this predicament; certainly Earth alone offers no prospect for our survival, let alone betterment, as a civilized species. The likely time between us and the spread of our descendants throughout the Galaxy is roughly comparable with the time between us and the Proconsul-ancestor of all apes and hominids. Given the novel intellectual, cultural and psychological demands of the new ocean, metabolic energy now spent on maintaining the heart, bones, and muscles against gravity, will become relatively surplus to requirements, and could eventually supply the raw material with which Creative Evolution will shape a more developed, Greater Man, or, indeed, several post-human species. These will carry on the torch of life as we picked it up from our simian ancestors. This is a distant possibility, but would lend our drive into space an added positive evolutionary significance.

Our children will not only be journeying to the final home, prepared for us long ago by the inscrutable order of the Cosmos, which many of us have known as God.

So far, we have considered this whole story from the view point of Humankind alone, but there are two other considerations. Firstly, the Earth is inhabited by several millions of species of creatures, derived from the same primaeval soup, and, in a larger sense, our kith and kin of the D.N.A double helix. These are of vastly differing evolutionary and cultural attainments, ranging from the primitive lichens, to the playful and intelligent dolphins. The ecological catastrophe which our activities threaten to bring about could seriously affect the future of other creatures on the planet. Our restless and expansionary nature is innate, and not to be readily suppressed; it will find its true outlet in the Humanization of space, and, at the end of the next few centuries, should leave behind a reduced human population, with much of human industry and commerce located elsewhere. This will give Mother Earth and her other children a respite from the imperial dominion of Humankind, and a chance to restore her battered ecology. We might truly see the Greening of the Earth-a prospect to delight environmentalists and space activists alike. Hence an early slogan of the Space Movement in America-"Conserve Earth-Colonize Space!"

Secondly, and finally, we shall, in the next few centuries make contact, whether by radio, archaeological discovery, or directly, with alien life or civilization-if such exists. Estimates vary widely as to the number of alien civilizations in our Galaxy; they range from 1.0 to 800 millions in our Galaxy at different levels. That being so, if we proceed in astronomy and space exploration , contact within the next century is highly probable; indeed, if there are any communicating civilizations within 100 light years of the Earth, contact is inevitable, since our radio and TV transmissions since 1945 have already radiated far out into space, and will be picked up in due course. Their reply, of course, will be an enigma; they may reply promptly, or they may have an ethos against administering culture shock to developing civilizations, and wait for us to grow up. Then, again, they may visit us. If they are advanced enough to do this, and were malevolent, it is likely that they would have already have destroyed us, so that I think it very unlikely that we would encounter actively hostile intelligence.

Fermi's Paradox famously asked "If there are so many Extra-Terrestrials, why are they not here?" One emerging answer could be that cosmic disasters destroy emerging civilizations before they achieve a Kardashev-2 civilization, or that advanced civilizations decay and give up the exploring and scientific culture which is necessary for the emergence of space based civilization. Thus, building a space-based civilization is in fact, as this book proposes, a direct measure of the life expectancy of a civilization. In terms of cosmic evolution, an "anti-Space" culture is a culture engaged in suicide. Or, then again, we may after all be the First.

Whatever form the encounter might take-two things are certain; one is, that, if there really are aliens, one day it will take place, and, secondly, that human history will never be the same again, for we will have ended, forever, the isolation of the cradle! Whether we put our present difficulties behind us, secure our future, and grow and develop in time to accept this ultimate challenge from the skies, as a mature civilization, or whether we slide into the overcrowded, internecine, impotent despair of a Hieronymus Bosch painting executed on a global scale-these are the choices before our generation. In the words of Arthur Kantrowitz,

"Children of Earth, go forth and conquer; you have a thousand worlds to gain, and nothing to lose but your Limits".

Chapter 11

The Space Movement.

For those who have read this far, and have come to share my conviction that a space-based civilization is the natural and logical next step for Humanity, there is good news. Many astronautical societies for enthusiasts, experts and activists alike are to be found throughout the the world. The International Astronautical Federation includes under its umbrella societies representing some 170,000 members , including the British Interplanetary Society. This Society was founded in 1933 and included H.G. Wells among its founders. its aims are to carry out research into astronautical and related topics, and to disseminate information on the achievements and benefits of space development in educational programmes and journals. Over the years, it has reached 3,500 members, of whom 1,000 are overseas. Many of these are employed in N.A.S.A, and have made major contributions to the space programmes in America, Europe, India, and Israel. The B.I.S headquarters is at 27-9 South Lambeth Road, London SW8 1SZ, and the Executive Secretary will supply information on request. They have produced classic pioneering papers over the years, including the 1939 project to send 3 men to the Moon, and the Daedalus Project, referred to earlier. Their 1982 Conference in Brighton in October 1982 was opened with cordial greetings from Lady Thatcher as well as Dr. Erik Quistgaard, then Director General of E.S.A. Their achievements in publicizing spaceflight were praised by Mr.V.A Kotolnikov, Chairman of the Intercosmos programme, and Mr. James Beggs of N.A.S.A. The Society publishes two principal Journals-Spaceflight for non-specialist readers, and the more technical Journal of the B.I.S. Either of these is indispensible for any space enthusiast who wishes to keep up with the news, and the annual subscription of $53 goes towards supporting a wide range of educational activities. At its headquarters is a very comprehensive library of astronomical and astronautical works.

The B.I.S is principally a learned society, with no real inclination to become either a mass movement like the Greens, or a political party. One of my aims in writing this book in the way I have is to arouse something more of an ideological/religious attitude among a larger section of the public towards space development, rather as the ecological Greens have done; although at first sight, Greens and space activism are rivals for the idealistic impulses of today's youth, they can be mutually supportive since the conservation of Earth by removing much of the most restless activity of Humankind off the planet serves both our ends without violating human freedom and dignity.

If sales of this work were to demonstrate a public enthusiasm towards space development , I propose to campaign for a programme aimed at putting Britain and Europe into the front line in humanizing space. I propose the following programme

1/ to commit E.S.A member governments to realize the four stage Lunar settlement plan to open up the Moon as described in this book :

2/ to start work, after a short FESTIP programme, on the construction of a SSTO capable of offering an airline type service to Low Orbit at $200 per kilo, with flights of 100 per annum initially, to be in service within 10 years of being pronounced technically feasible. More specifically, Skylon should be fully evaluated, and the relevant technologies demonstrated in full field trials; if Reaction Engines Ltd designs are further proven , then Skylon should be brought into production as a full and potentially superior competitor to the Lockheed Martin Venturestar

3/ a tax-free regime to be instituted forthwith for all Space-based activity by any E.S.A based enterprise, and

4/ a Declaration by the Member States, and the European Parliament, that the establishment of an extraterrestrial civilization is a concrete goal of European Civilization, with or without the collaboration of other powers.

The L5 International Space Society was founded in Tucson, Arizona in 1975 by Isaac Asimov, Keith and Carolyn Henson, Mark Hopkins, and Robert Heinlein who had become concerned with the run-down of the US Space programme after the Apollo and Skylab programmes in the early 1970's. The enthusiasm which took Apollo to the Moon was ebbing rapidly just as the true industrial potential of space was becoming understood by insiders. Indeed Vice-President Walter Mondale nearly succeeded in cancelling the Shuttle programme-a catastrophe which was only averted by the casting vote of President Carter. The public apathy and opposition to space ventures in the last generation has been largely due to ignorance among the ruling and educated groups as well as the man and woman in the street, about the real benefits of space development, and the larger ones to come. The L5 Society was formed largely to rectify this deficit but from the start was a more overtly campaigning group than BIS. It took its impetus from Professor O'Neill's work "The High Frontier" and has sought to make the Humanization of Space a popular goal, appealing to people in all walks of life.

In 1987, L5 merged with another group, the National Space Institute, to form the National Space Society, of 922 Pennsylvania Avenue, S.E.Washington DC 20003, USA. With a membership of over 25,000, and Chapters in every state of the U.S.A as well as several foreign countries, they are now a powerful voice in the land, and courted by politicians of all parties. Apart from this author's membership, the N.S.S includes on its Board of Governors several politicians in America, including ex-Speaker Newt Gingrich. His book, "Window of Opportunity", contains three chapters on the benefits to America of the coming revolutions in space and information technology, while Dr.Gingrich has attended a N.S.S Conference-I first encountered him in Houston in 1983, and can testify to his commitment to the development of space for the good of

America and all humanity. N.S.S members have addressed meetings, appeared on radio and TV shows, and contacted by direct mail tens of thousands of uncommitted people on both sides of the Atlantic. They have founded companies, testified at Congressional hearings, lobbied and educated Congressmen and Senators, and drummed up support at critical votes in the history of the Shuttle and Space Station programmes. The aim of disseminating information among the public and politicians is bearing fruit, with a gradual growth in the pro-space support in the US House of Representatives to over 50%, and the passage of Democrat George Brown's Space Colonization Bill, committing America to the goal of establishing a solar system-wide civilization as a national objective. In America, there is now a public majority for further advances in space, helped on by the link-ups with the Russians. During the 1996 Presidential election campaign, there were considerable lobbying efforts from the increasingly well-mobilized chapters of the National Space Society.

A TV teletext opinion poll in Britain in late1996 showed a 86% majority in favour of more UK involvement in space research, despite lower public awareness of the true potential of space as shown in this work. Space Shuttle launches are regularly watched by large crowds.

The Mars Pathfinder websites received 500 million "hits" in their first month of operation in July 1997-the most popular site ever in the Internet's short history. Membership of N.S.S supports campaigns in the U.S and Australia, most notably, as well as entitling the member to the quarterly magazine, Ad Astra. Many local radio stations turn to N.S.S activists regularly for commentary or information on space affairs-this is a valuable service to the media and public which can be performed by any reasonably well informed activist, and will help to sow the seed of public support for space activities. The N.S.S is of the opinion that there is a natural convergence of interest between environmentalism-except for those who wish to destroy human living standards and freedoms as a prelude to imposing a politically correct tyranny-and the search for world peace and justice, and the potential of the space movement. Not many people are aware of these possibilities unless they are put to them, and perception of these common interests can only advance the cause of space activism. The N.S.S now regularly acts as a public interest group on N.A.S.A / Congressional committees, and is increasingly heard as a major public voice on space matters.

For those who wish to support actual research into the development of space resources and technology, there is the Space Studies Institute, PO Box 82, Princeton, New Jersey 08542, USA. This was founded by Dr.Gerard O'Neill in 1977, after publication of "The High Frontier" in 1974, and the N.A.S.A-Ames sponsored study of space colonization in 1975-6, which concluded emphatically that the High Frontier scheme was technically feasible. In 1976-7 N.A.S.A reached a very low ebb financially, and so was unable to offer Dr.O'Neill any financial support for further research. In particular, they could not undertake studies of the Electromagnetic Mass Driver, and lunar chemistry experiments which are the lynch-pin of the whole programme envisaged by O'Neill. He and Dr. Freeman Dyson-now President of the Space Studies Institute-therefore decided to found the S.S.I, which is funded by a multitude of small grants from space enthusiasts, mostly in the scientific community; these also contribute expertise in research and publication, as well as time in speaking engagements and writings. The idea was that S.S.I would fund or carry out appropriate and focussed research on a long-term basis unaffected by the whims of politicians and public opinion, so that they could clarify the key ingredients for a successful High Frontier programme in the period 1977-1982.

This stage was successfully completed on time, and centered on the mass driver, lunar reaction chemistry, and mining, and design of Island One. The second phase, from 1982-7 produced detailed experimental models backed by working laboratory tests and prototype documentation of these ingredients, in time for a detailed, costed action plan to be put to Congress by 1987. By 1983, four main topics had been selected-mass driver, lunar reaction chemistry and mining, Solar power satellites studies based on maximum lunar material content , and suitable advanced automation techniques, to allow a massive increase in extraterrestrial industry for the humanization of space. The programme was completed, but the Challenger disaster has meant that the plan cannot be implemented in this decade. Research by the S.S.I has since focussed on reducing the size of the seed needed to start the programme, and into improving the detailed studies of lunar materials processing and S.P.S design. It may well be that a consortium of industry rather than government will eventually implement S.S.I's work. such a consortium could include S.S.I , or the space movement itself! S.S.I has tens of thousands of members contributing $15 per annum, and over 1,000 Senior Associates paying $100 per year; S.S.I publishes a bi-monthly News Update for its members. The feasibility of humanizing space in the O'Neill manner is now proven by hard data, and awaits only the will, and a more capable shuttle service. Another offshoot of S.S.I, launched by 3 Senior Associates in 1987, is the International Space University, which now runs degree courses in space-related subjects for a growing cadre of engineers, scientists, and administrators. Thus the ideas of S.S.I are finding their way increasingly into the training of the coming generation of aerospace planners. Links between S.S.I and a Russian equivalent organization, the Moscow Aviation Institute, have been forged over recent years, as with other American and Russian joint companies in the field of space activity. Much will be heard of these new developments in the next few years.

Students interested in undergraduate studies in space sciences and technologies can apply to the admissions Office of Kent University, Canterbury for details of courses in these areas; postgraduate study at Surrey University could lead directly on to employment opportunities in Small Satellites Technology Ltd, whose Managing Director Professor Martin Sweeting is possibly the world's most experienced designer and builder of small low cost satellites.

The Artemis Society (artemis@asi.org) aims to develop commercial space travel and lunar bases in the next 15-20 years by raising money from the media coverage of the programme in action. The programme would generate , it is believed, enough revenues from film rights and so on to fund itself, if it can raise the initial capital.

The Planetary Society, of 65 North Catalina Avenue, Pasadena ,CA 91106 tel 626-793-5100 http://planetary.org, was founded in the 1980's by the late Carl Sagan, and Bruce Murray of the Jet Propulsion Lab. and is the world's biggest space advocacy group, with some 110,000 members in many nations. They work to defend planetary missions against the ravages of political budget cutters, and also contribute to the design and operation of some missions

For those with access to the Internet, in 1996, I and some associates have formed an international campaigning group called Space Age Associates, at **http://www.astronist.demon.co.uk/index.html** to co-ordinate world-wide campaigning using writing, broadcasting, phone-ins and so on in order to raise public awareness of the opportunities before us. This group exists in cyberspace, is free of charge, and is administered entirely using the new medium. We now have members in the USA, Russia, Germany, Italy, Canada, Holland, Denmark, Israel and of course the UK. In October 1997, Space Age Associate has formed a loose collaboration with two likeminded groups-the "Foundation for the Scientific and Technological Research of Frontiers" (TDF) group, headed by Adriano Autino of Italy, who can be contacted at http://www.tdf.it , and the Orbiting United Ring Satellite (OURS) Foundation http://www.ours.ch based in Switzerland, with space artist Arthur Woods as President. Our three groups now have cross links, and share each other's material.

The First Millennial Foundation, founded by Marshall T Savage with the aim of initiating Space colonization from a base of extraterritorial marine colonies based on electricity generated from the temperature difference between deep and surface waters, and mari-culture

World Wide Web; http://www.millennial.org/ or E Mail rifle@ millennial.org UK branch E-Mail Address fmf-uk-info@trellis.co.uk

The Planetary Society-http://tps.planetary.org was founded by Carl Sagan and Bruce Murray, and numbers 120,000 members in over 70 countries.

Spaceguard UK, affiliated to the International Spaceguard Foundation, in honour of the Spaceguard described in Arthur Clarke's "Rendezvous with Rama", campaigns for greater awareness and research into the hazard posed by Earth crossing asteroids and comets, with a view to eventual prevention, and is led by many world famous scientists. It was founded in 1996 by Jonathan Tate at Jonathan TATE<fr77@dial pipex.com>

The address is:-
Spaceguard UK,
Cygnus Lodge
High Street, Figheldean, Wiltshire SP4 8JT
http://dspace/dial/pipex.com/town/terrace/fr77/
High Street Figheldean Wiltshire SP4 8JT
http://dspace/dial/pipex.com/town/terrace/fr77/

Finally, readers, I leave you on an optimistic but challenging note; the great opportunities of space lie within the reach of us and our children. Take them, and a brilliant civilization awaits our descendants; leave them, and our horizons will know only contraction and decay. In the final analysis, the choice lies with you, and millions like you.

Come and join us!

Epilogue

Birth of an Idea

"A startling new picture emerges. 70 million years ago, intelligence was, far from being a wildly improbable freak, advancing along three fronts towards three intelligent species in embryo-"Dinosaur Man", Mammalian Primate Man, and the Cetacea, and that all three were advancing towards consciousness before disaster struck in the shape of a large cosmic rock. The main goal of Creative Evolution is not mere D.N.A survival and replication-after all, that limited goal could survive even a succession of major collisions-but intelligence, and, in particular, Mind. Since the debacle of 65 million years ago Evolution has been working, using the available surviving material, to produce a species capable of doing the one thing that can ensure the survival and growth of Mind, as opposed to mere genes-namely, dispersal from this one planet before the laws of celestial mechanics interfere again, and thwart the whole enterprise. The above paragraph introduces a new idea of the human role in the universe-namely that of "Evolutionary Cosmic Destiny."

With the discovery, in March 1998 of lunar water ice, followed shortly afterwards by the worldwide banner headlines describing a possible major asteroid impact in 2028-it is likely that this Idea will gain ground in coming years, and may even lead to action in Space on a large scale. This idea was first set down by this author in his first draft for this book-in May 1984.

The fact that XF-11 is now expected to give us a decent berth affects the argument not one jot; we have seen the writing on the wall, and have been shown the way forward; it is up to us!

Bibliograph and Sources

Books;
Amateur Astronomy   ed. Colin Ronan   Hamlyn,London,1989
Ardrey, Robert :   African Genesis   William Collins,London,1972
Asimov, Isaac: A Choice of Catastrophes   Arrow Books, London 1981
                     The Collapsing Universe   Hutchinson,London,1977
                     Nemesis   Bantam, London, 1990
Bakker, Robert   The Dinosaur Heresies   Penguin Books, London 1988
Barrow, John, & Tipler, Frank   The Anthropic Cosmological Principle   Clarendon Press,Oxford,1986
Bauval, Robert & Gilbert, Adrian, The Orion Mystery   William Heinemann, London 1994
Berry, Adrian: The Next Five Hundred Years   Headline Books, 1995
Bova, Ben :   Colony   Magnum, New York 1979
Cavalli-Sforza, L&F The Great Human Diasporas Addison-Wesley Publ., New York 1995
Bonner, William: The Mystery of the Expanding Universe   Eyre & Spottiswoode, London, 1965
Brin, David: Earth   Orbit Books, London 1990
Bullock, John.& Morris, Harvey: Saddam's War   Faber and Faber, London,1991
Carlisle, David Brez   Dinosaurs, Diamonds, and Things from Outer Space   Stanford University Press, Stanford , California1995
Chaikin, Andrew& O'Leary, Brian & Beatty, Kelly   The New Solar System   Cambridge University Press, Cambridge, England,1982
Clarke, Arthur C.   Childhood's End   Pan Books, London 1956
                     Voices from the Sky   Victor Gollancz, London 1966
                     2001: A Space Odyssey   Arrow Books, London, 1967
                     Rendezvous with Rama   Pan Books,London,1974
                     2010: Odyssey Two   Granada, London, 1982
                     The View from Serendip   Pan, London 1979
                     The Sands of Mars   Signet Books, New York,1974
                     The Exploration of Space   Penguin, London, 1958
                     The Hammer of God   Orbit Sci-Fi, London 1994
Creighton, Michael   Jurassic Park   Arrow/Pan Books,London,1991
Daniken, Erich von: Chariots of the Gods   Corgi, London,1975
                     Return to the Stars   Corgi Books, 1974
                     The Gold of the Gods   Corgi 1975
De Chardin, Pierre Teilhard : The Phenomenon of Man   Collins, London, 1965
         Dyson, Freeman: Infinite in all Directions   Pelican Books,London,1989
The Epic of Gilgamesh trans. by Maureen Gallery Kovacs (Stanford: Stanford University Press, 1990)
Elegant, Robert: Mandarin   Hamish Hamilton,London,1983
Environmental Protection Agency, New York:
                     Report on the Greenhouse Effect of Fossil Fuels not known
Esslemont, E.J: Baha'u'llah and the New Era, Bahai Publishing Trust, London 1974
Evans, Richard: Deng Xiao - ping   Hamish Mamilton,London1993
Flenley, John & Bahn, Paul   Easter Island, Earth Island   Thames Hudson, London

1992
Fordham Frieda   An Introduction to Jung's Psychology,   Penguin, 1963
Gatland, Kenneth: Astronautics in the Sixties   Iliffe Books,London,1962
Gatland, Kenneth et al., Encyclopaedia of Space Technology, 1981 not known
Gibbon, Edward: The Decline and Fall of the Roman Empire, JMDent & Son, London,1986
Gilzin, Karl: Travel to Distant Worlds Foreign Language Publishing House
                                                    Moscow 1957
Gingrich, Newt: "The Millennium Project"- Speech delivered at Second Annual
                          Space Development Conference, Houston, Texas on 2/4/83
                          - sponsored by the L5 Society
Glubb, Sir John: The Empire of the Arabs   Hodder, London1982
Goldsmith, Donald   Supernova   Oxford UniversityPublications,1990
Gribbin, John: The Strangest Star   Fontana Books,London,1990
            In the Beginning   Penguin Science,London,1994
Haining, Peter: The Race for Mars   Comet/WHAllen, London, 1986
Half the World: Ed. Arnold Toynbee   Thames and Hudson, London,1973,
Henbest, Nigel., Marten, Michael   The New Astronomy   Cambridge U.P., Cambridge, England1985
Hampshire,Stuart: Spinoza   Penguin Books, London1981
Heyerdahl, Thor: The Ra Expeditions   Penguin, London,1972
Hirsh, Seymour: The Samson Option   Faber & Faber,London,1993
Horizon BBC TV: "The End of the Dinosaurs - examination of Luis Alvarez' Asteroid
                          impact theory" no details
Hoyle, Sir Fred :   The Intelligent Universe   Michael Joseph, London,1983
Humphreys, Christmas: Buddhism   Penguin Books,London,1981
Johnson, Paul: The Civilization of Egypt   Book Club Ltd, Weidenfeld & Nicholson, London,.1978
Kippenhahn, Rudolf 100 billion Suns   Counterpoint, Unwin paperbacks, London,1985
Koestler, Arthur,   The Heel of Achilles   Picador Books,London,1976
Kowal, Charles   Asteroids - their Nature and Utilization, Praxis Books, Chichester, 1996
Lancelyn-Green, Roger: Tales of the Greek Heroes   Puffin Books, London ,1961
Meadows, Dennis & Donella H. Meadows & Dennis L Randers & William Behrens III,
            Limits to Growth -   The Club of Rome. Earth Island, London 1972

Lovell, Sir Bernard: At the Centre of Immensities Hutchinson,London,1979
Lovelock, James   The Ages of Gaia, Oxford University Press,1990
with Michael Allaby   The Greening of Mars Andre Deutsch, London 1984
Lunan, Duncan: Man and the Planets   Ashgrove Press, Bath,1983
Mallowan,M.E.L   Early Mesopotamia and Iran   Thames & Hudson, London
Marshall, Robert: The Storm from the East   Penguin Books,London,1994
Meyers, Norman et al. Report of International Conference on Nuclear War in Washington, D.C
                          The Nuclear Winter - findings summarised in "Doomsday"
                          Pamphlet pub. Oct 1983 no details
Mitchener, James: Space   Secker & Warburg, London 1982

Moore, Patrick   Guide to the Moon   Lutterworth Press, London, 1976
The New Scientist   1983 Dec 17 Discovery of Fomalhaut System
                       1983 June Discovery of Vega System- no details
Muller, Richard   Nemesis - the Death Star   Mandarin Paperbacks, London, 1990
New Scientist Publications: Cosmology Today, IPC London, 1982
Nietzsche, Friedrich: Also Sprach Zarathustra, Penguin London,
                       Beyond Good and Evil, Penguin, London 1975
Nicholson, Iain: The Road to the Stars   Westbridge Books, London, 1984
Observing the Universe   ed. Nigel Henbest,   Basil Blackwell, Oxford, 1984

O'Leary, Brian: Project Space Station   Stackpole Books Harrisburg, 1983
O'Neill, Gerard K: The High Frontier Anchor Books, 1974 Doubleday, New York
                       2081   Simon & Schuster, New York, 1981
                       The Conquest of Space Article in Omni Oct 1983
                       SSI Updates 1983 Vol ix issues 1 and 2, p4 "Mass Driver
                       Update" Les Snively
                       SSI Update 1993, Vol xix, "High Frontier on the Threshold", Dr
                       Peter Glaser, pp1-4
The Penkovsky Papers   Doubleday & Co Inc., New York, 1968
Phillips, Kenneth: A Guide to the Sun   Cambridge University Press, Cambridge, England, 1992
Sagan, Carl & Schlovsky, Josef   Intelligent Life in the Universe
       Holden - Day, London, 1966

Sagan, Carl: Comet   Michael Joseph, London 1986
                       Cosmos, Macdonald/Futura, London 1981
Savage, Marshall T.   The Millennial Project; Little, Brown & co, New York, 1994
                       How to Colonize the Galaxy in   Eight Easy Stages
Schaeffer, Udo: The Imperishable Dominion   Gerald Ronald Pub.
                       46, Main St, Oxford, 1983
Shirer, William   The Rise and Fall of the Third Reich Pan Books, London, 1963
Stapledon, Olaf,   First and Last Men   Penguin London, 1968
Tayler, R.J : The Stars; their structure & evolution, Wykeham Science Series, Taylor Francis Ltd London, 1981
Temple, Robert: The Sirius Mystery Sidgewick & Jackson, London, 1976
Thomas P., & Chyba C., & McKay C   Comets & the Origin & Evolution of Life
       Springer, New York, 1997
Toffler, Alvin: The Third Wave   Pan Books, London, 1991
Verne, Jules: Round The Moon   Airmont Publishing New York, 1969
A Void Home   Video by Laserradio
Wendt, Herbert   Before the Deluge   Gollancz, London, 1968

Will the Wells run Dry   Royal United Services Institute 1978 report, British
Atlantic   committee, RUSI building, Whitehall SW1A 2ET
World War Three   ed. Shelford Bidwell   Hamlyn, London 1979
Zubrin, Robert   The Case for Mars, Simon & Schuster, New York, 1979

JOURNALS

Ad Astra — National Space Society, bimonthly

Amateur Astronomy & Earth Sciences/ Quest for Knowledge magazines Jan 1996 to Jan 1998 articles by this author

Ashford, D.M.,   Spacecab 2: a small Shuttle   )   Spaceflight Jan.1985 vol27, number 1

     Space Tourism: an Update   )   Spaceflight Mar1996 vol35 number 3

Astronomy   Kalmbach Pub. Corp, Milwaukee, Wisconsin

Fearn, D.G, Gray.H., and Smith.P., UK Ion Propulsion system development, J.B.I.S, vol49 number 5 May 1996,pp162-173

Spaceflight, and Journal of the B.I.S, publications of the British Interplanetary Society
27-9 South Lambeth Rd, London,SW8 1SZ

Soviet Year in Space, edited by Nicholas Johnson, Teledyne Brown Engineering,Colorado1986-90

L5 News - the External Tank - the road to Space - David Brin 1983 no details

SSI Update   Space Studies Institute Newsletter, quarterly

SSI Updates 1983 Vol ix issues 1 and 2 , p4 "Mass Driver Update" Les Snively

SSI Update 1993, Vol xix, number 2, "High Frontier on the Threshold", Dr Peter Glaser, pp1-4

Space Biology-Workshop at DFVLR Institute of Aerospace Medicine, Cologne 1983, sponsored by European Space Agency, and European Low Gravity Research Association no details

Columbus logbook, Magazine of the European Space Research Organization, pub. quarterly in Holland no details

Space, the Crucial Frontier-Citizen Advisory Council on National Space Policy 1981, Jerry Pournelle, Chairman. Edited by L5 Society

Project Daedalus-Journal British Interplanetary Society., London 1981

Contributed works

New Trends in Astronomy Teaching a Colloquium of the IAU held in London 1996,

ed L. Gougenheim, D. McNally, J.R. Price, Desktop Space Exploration, M.Martin-Smith and R.A. Buckland, Cambridge University Press, Cambridge 1988, pp237-41

Acta Astronautica, 1998 vol 43 (issue8 pp243-46) Desktop Space Exploration M.Martin-Smith and R.A. Buckland

Journal of Space Policy Nov 1998 vol14, number 11, pp133-5, Elsevier Oxford, A Historical Perspective on Space, M. Martin-Smith

Spaceflight pub BIS London April 1997, "Into Space", vol 38 issue 4, p142 feature M. Martin-Smith